Don Hinrichsen is a London-based writer, specializing in environment and development issues. He is the European correspondent for *Amicus Journal* in New York and Contributing Editor to *Earthwatch*, the population-environment supplement of *People* magazine in London. Don Hinrichsen is also a consultant to various United Nations agencies, in particular the UN Population Fund in New York. Prior to this, he was Editor-in-Chief of *Ambio*, the international journal of the human environment published by the Royal Swedish Academy of Sciences in Stockholm; and Editor-in-Chief of the *World Resources Report* published by the World Resources Institute and the International Institute for Environment and Development. He has also worked as a foreign correspondent for radio, newspapers and magazines. His previous books include *Our Common Future: A Reader's Guide* (Earthscan, 1987).

Our Common Seas: Coasts in Crisis

Don Hinrichsen

EARTHSCAN PUBLICATIONS LTD London
in association with
United Nations Environment Programme, Nairobi

First published in Great Britain 1990
by Earthscan Publications Ltd
3 Endsleigh Street, London WC1H 0DD

British Library Cataloguing in Publication Data
Hinrichsen, Don
Our common seas.
1. Oceans. Environment. Conservation
I. Title
333.916416
ISBN 1–85383–030–5

Typeset by Selectmove Ltd, London
Printed and bound in Great Britain by Cox & Wyman Ltd, Reading
Production by David Williams Associates (081–521 4130)

Earthscan Publications Limited is an editorially
independent and wholly owned subsidiary of the
International Institute for Environment and Development (IIED)

Contents

Acknowledgements

Despite long nights and the rewrites, I want to thank Richard Sandbrook, Director of the International Institute for Environment and Development (IIED), for submitting my initial book proposal to the United Nations Environment Programme (UNEP) for funding. And, of course, I want to express my gratitude to Stjepan Keckes, Director of UNEP's Oceans and Coastal Areas Programme Activity Centre in Nairobi for agreeing to fund such a daunting project.

The research for this book took up a full six months, including four months of extensive and exhausting travelling. In the process I was able to visit most of UNEP's regional seas programmes, crossing all the world's seas except the east coast of South America, the polar seas and the sea areas of northern North America and Europe (as they are outside the context of this book). Unfortunately, I was unable to include all of the world's regional seas in my analysis.

In any case, this book, like all books, would not have been possible without encouragement, advice, help and prodding from a host of people. I cannot list them all but the following people deserve special mention: Lloyd Timberlake of IIED, for reading early versions of the manuscript; the entire staff of UNEP's OCA/PAC office in Nairobi for reviewing the regional seas chapters, especially Arthur Dahl, Deputy Director of OCA/PAC, for his critical comments and suggestions for improvements, and Reza Amini for general moral support; Drs Ed Gomez, Helen Yap, and John and Liana McManus from the Marine Science Institute of the University of the Philippines in Manila for excellent interviews and for co-ordinating my programme while in the Philippines; Dr Angel Alcala, Director of the Marine Laboratory at Silliman University, the Philippines for hosting my stay on the island of Negros, and Louella Dolar, a faculty member in the Biology Department of Silliman University for giving valuable time from her teaching duties to accompany me around Negros, often acting as an interpreter; Jeannie Peterson, Director of UNFPA's Philippine office for providing me with a car and driver while in Manila (I would never have made any of my appointments without one); Winarta Adusibrata, old friend and colleague, who accompanied me to the fishing village of Muara Angke, north of Jakarta, and acted as an impromptu interpreter and guide; Jairo Escobar, UNEP consultant to the Permanent Commission for the Southeast Pacific in Bogota, for co-ordinating my far-flung and haphazard travels in Colombia, and to his entire staff who helped translate material for me from Spanish into English and assisted in many small ways to make my stay productive; Hortensia Broce, formerly with the Panamian National Environment Commission, for acting as guide and interpreter while I travelled around Panama; Sebia Hawkins and her Greenpeace colleagues in Washington, DC, for their comments on the Pacific chapter; Dr John Pernetta and Dr David Mowbray of the University of Papua New Guinea, for critical reviews of the Pacific chapter; Dr Badria Al-Awadi, co-ordinator of the Regional Organization for the Protection of the Marine Environment (ROPME), for organizing my programme in Kuwait; and James Bishop, fellow ex-patriot who gave me valuable insight into the Arab world. Finally, I would like to thank those people who shared their lives with me and helped me to understand their problems and experience them first hand.

Don Hinrichsen,
London
December 1989

Preface

The 1990 Report on the State of the Marine Environment* highlights the rapid deterioration in the coastal areas of the world's oceans, which threatens to cause significant harm to the marine environment in the next decade unless strong, co-ordinated national and international action is taken now.

In order to bring the crisis affecting coastal areas to the widest possible audience, the United Nations Environment Programme (UNEP) supported the International Institute for Environment and Development (IIED) in preparing a book describing what is happening to coastal areas around the world, with particular reference to the coastlines covered by UNEP-sponsored Regional Seas Action Plans, and documenting the efforts being made to solve the problems. This book by Don Hinrichsen is the result. It should be emphasized that the views expressed herein are those of the author and may not necessarily correspond with the views of UNEP or IIED. Similarly, the designations employed and the presentation of the material do not imply the expression of any opinion whatsoever on the part of UNEP or IIED concerning the legal status of any State, Territory, city or area, or of its authorities, or concerning the delimitation of its frontiers or boundaries.

If parts of this book make for depressing reading, it is because the problems are growing rapidly, and the efforts being made to control them still far from adequate. Hopefully it will inspire those who read it to increase the pressure for effective national and international action to protect and manage our fragile and critically important coastal areas while there is yet time.

*GESAMP (IMO/FAO/Unesco/WMO/WHO/IAEA/UN/UNEP Joint Group of Experts on the Scientific Aspects of Marine Pollution): *The State of the Marine Environment*. UNEP Regional Seas Reports and Studies, No. 115. UNEP, 1990.

Introduction

The world's coastlines, mercuric and ever-changing, have been settled since the dawn of time. They have nurtured humanity through countless centuries.

Today coastal areas are beginning to sag under a human onslaught. Home to much of the world's population, coasts and near-shore areas are being over-run with people. Six out of ten inhabitants of the "blue planet" now live on or within 100 kilometres of the seashore. The result is too much development and too much competition for limited resources. Coastal cities and towns multiply out of control, like some genetic experiment gone awry. Industries spread with little regard for environmental amenities. Resources are used and abused with thoughtless abandon. Everywhere, it seems, coastal zones are under tremendous pressures.

Despite the threats, the resources of the world ocean are regarded as infinite. This homocentric notion is reinforced by the vastness of the world's seas, which cover nearly 362 million square kilometres. The earth is really misnamed. It should be called "water", or more accurately, "sea". As the first satellite pictures of earth dramatically demonstrated, our planet is 70 per cent seawater. Viewed from space, the main feature of earth is its vast interconnected ocean. Land appears almost as a geological after-thought. The Pacific Ocean alone accounts for almost 32 per cent of the planet's surface, more than all the land masses put together. Yet we know precious little about our seas, or how they work. The biology of most land creatures is well-known, as are biogeochemical cycles. Yet we know little about many of the ocean's residents, or how they interact. Except for general patterns, not much is known about currents – those great underwater highways constantly on the move.

What we are learning – the hard way – is that our seas are not immune from what happens on land. Much of humanity's refuse ends up sooner or later in coastal waters, washed down by rivers or dumped directly. Rivers also bring into the seas billions of tonnes of sediment. Bound to these sediments are unknown quantities of poisonous residues from agricultural chemicals and heavy metals from industrial discharges.

Ocean pollution is now universal. Toxic chemicals – such as pesticide residues – dumped in the North Atlantic may very well end up in the South Atlantic or the Antarctic. Radioactive waste tossed into the South Pacific may find its way into the food chain of Indian Ocean fisheries, or even that of the Arctic. DDT residues were found embedded in the fat of Antarctic penguins and seals, thousands of kilometres away from any potential source. Lead compounds were discovered in open ocean surface layers in the Atlantic. Polychlorinated Biphenols (PCBs) were found in sediment taken from under the Arctic icepack. Plastic pollution is now a blight – and a deadly menace – for marine life on all our seas.

Meanwhile, exploitation of coastal resources – mangrove forests, estuaries, seagrass beds, coral reefs, and the fisheries they foster – has reached an unprecedented scale. Even open ocean fisheries, once thought inexhaustible, are now being fished to their limits. Some fisheries, like herring and cod, have been periodically overharvested in the North Atlantic and in the North and Baltic Seas. Others, such as the Peruvian anchovy fishery, have collapsed entirely, due to a combination of over-exploitation and the bizarre weather patterns known as *El Niño*.

It is high time to start managing the world's seas and coastal areas, instead of merely exploiting them. Although the crisis of our coasts has been recognized by some governments, international agencies (like the United Nations Environment Programme and UNESCO) and non-governmental organizations (like Greenpeace), we have a long way to go. Much of what we know about the state of our seas is due to the collective work of thousands of over-worked and under-funded university research centres and other bodies set up to study various aspects of marine science. For the most part, high-powered think-tanks have yet to discover the sea.

Despite all the bad news, there are signs that more and more governments – both North and South – are beginning to recognize the value of managing their coasts and near-shore waters. Coastal area management plans are being drawn up by over 50 countries, while 20 others have management regimes already in place and working. International co-operation in pinpointing and solving common problems related to the use and abuse of shared seas is on the increase. The international aid community is also shaking itself awake to the enormous problems posed by the over-use and destruction of coastal resources.

One United Nations organization has been a catalyst in spurring on scientific research and international co-operation on solving the crisis of our coasts. The United Nations Environment Programme's "Oceans and Coastal Areas Programme Activity Centre" has been instrumental in ringing the alarm bells of concern and in getting sovereign governments, sharing a common sea, to deal with their problems in concert. From a modest beginning in the mid-1970s, with the launching of the Mediterranean Action Plan (signed by all Mediterranean countries, except Albania), UNEP went on to set up nine more "regional seas programmes" involving 120 countries.

Although shocking, the alarming contents of parts of this book are not meant just for shock value. The problems described should not be underestimated. But at the same time, behind the problems are people and institutions trying to make a difference. People, after all, are both the cause and the cure. Most of the abuse heaped upon earth is human-induced. The buck stops with us.

More than a portrait of pollution, this book is also a portrait of people – from the artisanal fisherman in the Philippines to the Greek ship-owner – who are dependent upon and who draw their livelihoods from the world's seas. And it is a portrait of what the collective action of communities, governments and international agencies can accomplish.

No book yet written on the world's seas has tackled all the issues related to such a vast and complex topic. This book does not even attempt to come close. By concentrating on the seas of the South, it is limited in scope to the problems and prospects of tropical and sub-tropical developing countries (with the exception of most countries rimming the Mediterranean). As with land-bound development issues like the destruction of tropical rainforests, the crisis of the coasts is most pronounced in those areas lacking the personnel, money and motivation to do something about them. This focus of the book, however, in no way is intended to exempt the North from a great deal of responsibility for the mess our seas are in. Not one developed country with a coastline is blameless. But books must begin and end somewhere. This one concentrates on that wide swathe of ocean around which two-thirds of humanity lives, and where people are increasingly suffering from the limitations of resources.

Our seas are embroiled in local, regional, and international conflict. Just being aware of what is at stake is a beginning. More important, however, is the next step: getting people to do something about it. Although much of this book is devoted to the crisis of our coasts, it is dedicated to all those who are struggling – against great odds – to make a difference.

1. On Distant Shores

In the year 2100, assuming we last that long, history may very well judge the late twentieth century by its catalogue of failures, rather than its successes. True enough, we do live in extraordinary times. Peace is breaking out all over the world. Traditional East–West rivalries are being discarded. Co-operation is replacing confrontation. The Cold War may soon be confined to the dustbin of history. But all these wonderful political, economic and social achievements may be overshadowed by humanity's frustrating inability to manage sustainably our resource base in accordance with the needs of a growing global family. A healthy environment is not a luxury, it is an absolute necessity if life on this planet is to be sustained into the twenty-first century and beyond.

The decade of the 1990s is being called the "decisive decade". Decisions not taken over the course of the next ten years to resolve the critical resource and population challenges facing humankind will no doubt undermine the future of generations born next century. Already we are losing a minimum of 6 million hectares of agricultural land to erosion every year. Desertification is threatening about one-third of the world's land surface, or 48 million square kilometres. Tropical rainforests, repositories of immense biological wealth, are being destroyed at the rate of 10 million hectares a year – an area the size of Austria. Through ignorance, greed and neglect, we may be condemning several million species of plants and animals to extinction.

In some cases it is too late to alter biogeochemical processes that have already been set in motion. It appears that humankind is facing the almost certain prospect of global climate change on some level, regardless of what we do during the next decade. Because of the great quantities of carbon dioxide and other trace gases produced by industries and vehicles the atmosphere is heating up like a greenhouse. But if nothing is done to ameliorate rising sea levels and hotter temperatures, the effects of widespread climatic change could have devastating consequences for developed and developing economies alike. Instead of confronting a crisis that could be contained, if we acted in time, we may be faced with a human and resource tragedy of untold proportions.

Indeed, in many regions throughout the world, humankind appears to be waging a kind of war against coastal ecosystems and marine life. Marine mammals – especially seals and dolphins – are killed off by the hundreds of thousands, entangled and drowned in fishing nets or deliberately wiped out because they compete with fishermen for fish. Fish and shellfish succumb to lethal doses of pollution flushed into coastal waters by industries and municipalities. Increasingly, seas are used as sewers and garbage dumps. Coastlines are fouled with all manner of industrial detritus and discarded plastic. Coastal habitats, such as wetlands, are smothered by imported sediment torn out of agricultural lands in the interior. Mangroves, estuaries, seagrass beds and coral reefs are plundered in the name of development.

Although coastlines everywhere are suffering from an influx of people, accompanied by resource destruction, the coastal crisis is qualitatively and quantitatively different between North and South. Coastal zones in the developed, northern countries suffer more from a glut of untreated industrial and municipal wastes, than from indirect effects of runaway population growth. Furthermore, the rich countries have the capital and technology to control pollution of coastal waters. They have a wide array of policy options and technologies to choose from. For example, turning off the pollution pipeline would make a big difference in the North and Baltic Seas.

For the South, however, there are fewer options, and fewer solutions. Third World coastal nations have the unenviable task not only of coming to grips with pollution, but they must also try to cope at the same time with the human and resource consequences of burgeoning coastal populations. In many instances grappling with the population challenge means dealing with the causes and consequences of migration from the hinterlands. It means developing strategies to attract people away from coastal areas, especially cities and towns. It means confronting a host of complicated, but related, problems, such as finding ways to keep upland farmsteads viable and productive. There are no easy technical fixes for the developing world's coastal crisis.

For these reasons, the focus of this book is on the seas of the South, and particularly those regions with the highest rates of population growth and the least amount of resources, on a *per capita* basis, to deal with this growth. Year by year, as numbers escalate, resources shrink and traditional management systems begin to break down. As with many other environmental and resource crises, developing countries are least equipped to cope with either the causes or the consequences of coastal degradation.

People and their needs are at the centre of the problems and at the centre of the solutions. But as coastal populations continue to grow, solutions become harder to find and to implement. Along the over-populated coastlines of Pakistan, India, Bangladesh and Sri Lanka, for example, population densities often reach up to 500 per sq km, more than twice the number of people living in the interior. On the island nations of the Philippines and Indonesia, coastal populations are growing by 4 per cent or more a year, compared to rates of 3 per cent or less for the uplands. Probably 60 per cent of humanity, or nearly 3 billion people, live on or within 100 km of a sea coast. In parts of Southeast Asia, 75 per cent of the population lives along the coast.

Human numbers overwhelming coastal resources and ruining marine habitats are certainly not difficult to perceive. Take the Mediterranean Sea. Any population density map shows the problem in graphic detail. The entire coastline is etched in with a black ring – thicker on the northern rim than the southern – representing coastal crowding. A satellite photo at night would reveal a necklace of lights around the entire seacoast. From space, it looks like a continuous coastal village. Indeed, demographers are predicting that the Mediterranean's resident population may swell to 200 million by the beginning of the next century.

But there is another, often overlooked, aspect to coastal degradation and pollution; one that is not so easy to see, or to combat. Coastlines also pay the ultimate price for mismanagement of the *land*. Watersheds denuded of forests quickly erode. Sediment is washed out of the hills and transported to the coasts via rivers and streams, where it smothers estuaries and coral reefs, depleting fisheries. The River Ganges, which drains an area of just over one million sq km of intensively-worked farmland and heavily-harvested forestland, delivers 1.46 billion tonnes of sediment to the Bay of Bengal every year. By contrast, the Amazon River, which drains a much larger area, takes only 363 million tonnes of sediment into the Atlantic each year. Runoff from fertilizers and pesticides poisons freshwater systems, which eventually dump their toxic load into coastal waters. Wastes discharged into rivers eventually reach the sea.

Attempts to make the critical links between land and sea often flounder on the shoals of development and economic priorities that sacrifice one for the other. Integrated management of land and sea has been beyond the capacities of most governments. Even developed countries – with a few exceptions like Sweden – have not managed to evolve integrated management strategies. Coastal zones may be managed, but such management is often divorced from what happens a few tens of kilometres inland. In much of the Third World there is no resource management at all.

Coastal resources in danger

The loss of vital coastal resources – mangroves, seagrasses, coral reefs – can be attributed as much to ignorance as to outright greed. In many areas of the world, coastal peoples do not understand the crucial role these ecosystems play in providing them with food, fibre, building materials and other necessities. Conservation is not usually something that comes easily to the poorest billion people living on the edge of survival.

Sustainable management is impossible if coastal communities are forced to live from day to day. It is also impossible if the consumption of the richest billion is not curbed. For example the Japanese woodchip industry destroys more mangrove forests in Indonesia than poor fishing communities who chop them down to build fish ponds. Still, the combined effects often obliterate mangrove swamps. With mangroves gone, fish and shellfish catches decline, robbing those who fish of their livelihoods and robbing populations of essential sources of protein. Incredible as it may sound, the importance of mangroves, seagrasses and coral reefs to the human food chain is not recognized by many small-scale and subsistence fishing communities. They have evolved many ingenious ways to get food, but they have no idea where the fish come from, how they reproduce, or what factors make fish stocks increase or decrease.

The importance of coastal wetlands as spawning, nursery and feeding areas for fish and shellfish cannot be understated. Globally, nearly two-thirds of all fish caught are hatched in tidal areas. Roughly 90 per cent of all commercial species of fish caught in the Gulf of Mexico and the Caribbean are dependent on estuaries, mangroves, seagrasses and coral reefs for critical stages in their life cycles. Likewise, some 80 per cent of the Indian fish catch from the lower delta region of the Ganges and Brahmaputra Rivers comes from the mangrove swamps of the Sundarbans, which cover 6,000 sq km. Half the current 500,000 tonne take of fish in the Lower Mekong Basin is of wetland origin. In 1981, the fisheries value of the Mekong Delta contributed around $90 million to the struggling economies of Cambodia, Thailand and Vietnam, while supplying 50–70 per cent of the protein needs of the Delta's 20 million people. Tropical Pacific islanders depend on coastal marine resources for at least 90 per cent of their protein intake.

Biologists consider mangrove forests to be one of the most productive and biologically diverse wetlands on earth, supplying important habitats for over 2,000 species of fish, invertebrates and epiphytic plants. The root zones of mangroves provide sanctuary for sponges, crested worms, crustaceans and molluscs, as well as green, red and brown algae. Intertidal zones create habitats for a variety of crabs and small animals, while hundreds of species of birds nest in mangrove canopies. Mangrove estuaries shelter marine mammals such as dugongs, manatees and otters, as well as reptiles like the endangered South American caiman and the Indo-Pacific crocodile.

Some 60 species of salt-tolerant mangrove trees and shrubs cover roughly 24 million hectares of intertidal, lagoonal and riverine flatlands throughout the world. Most species are found in a wide tropical belt, reaching their greatest concentrations along the coasts of South and Southeast Asia, South America and Africa. The largest expanse of mangrove forests – about 20 per cent of the world's total – borders the Sunda Shelf, a region in Southeast Asia encompassed by Vietnam, Kampuchea, Thailand, Malaysia and the Indonesian islands of Sumatra, Java and Borneo.

Mangrove communities manufacture a nutrient-rich broth for sustaining a wealth of marine life. Concocted from the decomposition of mangrove leaves and twigs, this broth is the first link in a long food chain that extends seaward through seagrass meadows, to coral reefs, and finally to open ocean fisheries. The abundance of offshore shrimp, for example, is directly related to the amount of mangrove nurseries available. On the Fiji Islands, about half of all fish caught by commercial and artisanal fishermen are dependent on mangrove forests for at least one stage in their life history.

Scientists have estimated that one hectare of mangrove forest, if properly managed, could produce an annual yield of 100 kilogrammes of fish, 25 kg of shrimp, 15 kg of crabmeat, 200 kg of molluscs, and 40 kg of sea cucumber. In addition, this same area could supply an indirect harvest of up to 400 kg of fish and 75 kg of shrimp that mature elsewhere.

Unfortunately, properly managed mangroves are a rarity. In many areas of the Third World, they are not managed at all. Despite their value as fish and shellfish nurseries, land builders and shore stabilizers, and wildlife habitat, mangroves are in retreat throughout their range. Clear-cutting for timber, fuelwood and wood chips, and outright destruction for the creation of brackish fish and shellfish ponds, and for the expansion of urban areas and agricultural lands has claimed millions of hectares globally. Indirect threats include run-off from pesticides used on agricultural fields and erosion sediment brought in by coastal development, the deforestation of upland watersheds and mining operations.

In the everyday struggle for survival it is easy for hard-pressed coastal communities to overlook the importance of mangroves, until it is too late. The pace of mangrove destruction rivals that of tropical forests. Yet there is a stupefying lack of awareness of the problem and its consequences. From government policy-makers down to subsistence fishermen, mangrove forests are often regarded as wastelands, to be exploited and converted to more "productive" uses.

In many cases the damage is not irreversible. Small-scale mangrove rehabilitation projects have been started up in fishing communities in South and Southeast Asia, Africa and the Caribbean. Much remains to be done, however, before the destructive trends of the last 40 years can be halted and reversed. Education is the first step. Without an appreciation for the value of mangroves, particularly among commercial fish pond operators, timber harvesters and artisanal fishing communities, the destruction is likely to continue.

Much of the damage being done to coral reefs, however, may prove very difficult to reverse. Of all the vital coastal ecosystems under threat, it is coral reefs – the marine versions of tropical rainforests – which are being decimated faster than any other marine resource. It is possible that they are being extinguished more rapidly than rainforests.

Although there are estimated to be 600,000 sq km of coral reefs in the world, with 30 per cent concentrated in Southeast Asian seas, their condition is thought to be deteriorating nearly everywhere. Wherever coral reef inventories were carried out, the results have been depressing. Of 632 reefs surveyed in the Philippines, for example, only 10 per cent were found to be completely undamaged.

In an extensive coral reef survey sponsored by the United

Nations Environment Programme and the International Union for Conservation of Nature and Natural Resources (IUCN), marine biologist Sue Wells discovered that dynamite fishing is a serious problem in 41 countries, while 50 countries report excessive sedimentation on parts of their reefs.

The main threats to reefs include a combination of improper land management, especially deforestation of coasts and uplands, and outright destruction through coral mining and blast fishing. As forests are cut down for timber and to make room for more agricultural land, huge quantities of erosion sediment can be flushed off the land and into shallow coastal waters. Without mangrove forests to trap the sediment, it is often transported to coral colonies, where it suffocates the living polyps. Sedimentation of reefs can transform thriving communities into dead ones. When forests are converted to farmland, the run-off of nutrients is greatly accelerated. Nitrogen and phosphorus poison corals and reduce their ability to compete with other organisms, such as algae.

Coral mining and blast fishing are particularly destructive since most coral species grow very slowly. It takes about 20 years for a brain coral colony to grow as big as a man's head. Coral reefs are like oases in the desert; once destroyed, the habitat reverts to "desert", where only a tiny fraction of their former inhabitants can survive.

The destruction is made more horrific when one considers that coral reefs are among the oldest living communities of plants and animals on earth, having evolved some 450 million years ago. Most modern coral reefs are between 5,000 and 10,000 years old; many of them forming thin veneers over older, much thicker reef structures dating back several million years. Most of the reef colony is actually dead. Only the upper layer is covered by a thin changeable living skin of coral. Coral polyps (the animals that build reefs) are the master bricklayers of the sea, cementing their homes upon the remains of their predecessors.

Coral reefs rival tropical rainforests in species richness and diversity. The 150-km long barrier-reef surrounding Palau in the Pacific, for example, has nine species of seagrasses, more than 300 species of corals, and 2,000 varieties of fish. The Great Barrier Reef of Australia has 400 species of coral providing habitat for over 1,500 species of fish and 4,000 different kinds of molluscs.

Nearly one-third of all fish species live on coral reefs, while others are dependent on reefs and seagrass beds for various stages in their life cycles. Although estimates as to the amount of fish that can be harvested from reefs varies from 1–30 tonnes per sq km per year, a sustainable harvest of some 15 tonnes per sq km of fish, molluscs, and crustaceans should be possible. Almost 90 per cent of all fish caught by artisanal fishermen in Indonesia are reef-dependent, as are some 55 per cent of the fish consumed by Filipinos.

The high productivity of coral reefs is due to their efficient biological recycling and retention of nutrients. Biologist Rodney Salm, Project leader of IUCN's coastal-zone management programme in Oman, explains how this remarkable recycling system works:

> The coral animals have tiny algal cells, called *zooxanthellae*, in their tissues. These process the coral polyp's wastes before they are excreted, thereby retaining valuable nutrients. Nutrients such as nitrates, phosphates and carbon dioxide produced in the polyp are used by the *zooxanthellae* during photosynthesis to generate oxygen and organic compounds which in turn may be used by the coral polyp. In this way the *zooxanthellae* recycle waste products to form nutrients within the tissues of the coral animal, saving the polyp the energy it would have expended on these activities. In addition, movement of water over reefs by waves and currents constantly washes the corals, and eliminates the need for the polyps to clean themselves. The energy freed up in these ways can go into damage control and new growth.

Reef-building coral animals, efficient as they are, still depend on sunlight. They cannot reproduce in murky water. Unless the water is exceptionally clear, most coral growth stops at a depth of 20 metres (although coral growth as deep as 40 has been found).

In order to appreciate the wealth of marine life that reefs shelter, coral communities can be viewed as "apartment house complexes", where many different kinds of organisms find shelter and food. The secret to the reef's richness of species lies in its complex architecture. Resident reef fishes, which forage during the day, share their living quarters with other, nocturnal species. While the diurnal (daytime) fishes are feeding, their living space is occupied by a fish which is only active at night. During the night the roles are reversed. This sharing of quarters allows a reef to shelter two separate populations of fish. It has been suggested that reefs can support 5–15 times the number of fish found in the North Atlantic. "But putting yields aside," points out Rodney Salm, "the fundamental thing to remember is that coral reefs are no-cost, self-perpetuating fish farms which produce high quality protein from essentially empty sea water."

The high species diversity on coral reefs gives rise to another, often over-looked, benefit: their potential as sources for new drugs. In order to help reef organisms cope with competition, many have developed substances harmful to other organisms. Researchers have discovered that a number of these highly active compounds may have useful medical applications. For example, certain reef-dwelling seafans and anemones possess compounds with antimicrobial, antileukemic, anticoagulant and cardioactive properties. Such species may prove invaluable in developing anti-cancer drugs, or other pharmaceuticals. The Australian Institute of Marine Science, for instance, has isolated a compound which protects the coral from sunburn. This compound has great potential for application in sunscreen products, and is in the final stages of testing for commercial production.

Coastal zone management

There are no easy solutions to the human and resource crises afflicting the world's coastal areas. Land and sea must both be managed in a way that permits economic development yet sustains the resource base. This involves the balancing of a multitude of human uses with each other, as well as managing resources in such a way that future needs are not sacrificed for the expediency of the moment.

Most coastal management schemes, unless painstakingly worked out with coastal communities and their needs in mind, soon find that they worsen problems they were meant to solve. Tanzania learned this lesson the hard way. In the late 1960s, the government decided to set aside some marine reserves south of Dar es Salaam. No one bothered to try and build support for the reserves among the local fishing communities. Several years later when researchers visited the proposed sites to take an inventory, they were appalled when five of the reefs slated for protection could not even be found. Villagers had blasted them apart for building material.

The other blatant shortcoming of most management schemes is that they are not integrated with land management or coastal resource-use. Fisheries management is nearly always that and nothing more. Caretaking mangrove forests, seagrass beds and coral reefs, upon which most tropical fisheries are dependent, is not included in traditional "fisheries management". And there is little co-operation between relevant government departments. Mangroves are cleared for agricultural land under the jurisdiction of an agricultural ministry which does not take account of the decline in fish catches that the loss of mangroves entails. Fisheries suffer, while agricultural output increases. One group benefits, another is impoverished.

And in the long term, often no one benefits at all. When mangrove swamps were cleared for agricultural plots on Fiji, the improperly drained soils soon became too acidic to support rice. After a few harvests, the land was abandoned. Only an expensive reclamation scheme could salvage the soil for crop production, or anything else.

Tourist development, without the benefit of coastal management plans or environmental impact studies, can be absolute folly. An example is Dumaguete City on the island of Negros (the Philippines), which decided to lengthen the runway at the local airport so that bigger jets could land with more tourists. But planners failed to take account of the environmental impacts that might result from extending the runway into the bay. No one paid any attention to likely impacts on the city's coastal areas. A few months after the runway was completed, the resort hotels along the coast noticed that their beaches were eroding. Huge wedges of sand were being gouged out and carried away. Subsequent studies revealed the cause: extending the runway into the bay altered tidal currents, which now sweep along the shore scouring out the beaches. A hasty wall had to be built so that the hotels themselves would not be devoured by the sea. Since tourists seek clear water and generous beaches, few bother to come to Dumaguete City.

Trying to close off coastal environments to any kind of development is pointless, unless proper management is introduced. Hundreds of marine parks and protected areas have been declared throughout the world, but many of them suffer from the lack of even a simple management plan. When the Organization of American States (OAS) carried out a review of protected marine areas in the Caribbean in 1986, they discovered that there were 112 with legal protection, accounting for nearly a quarter of the protected marine areas in the world. On closer inspection, the OAS research team found that of these 112 marine parks and reserves, only 28 of them had a budget and staff, a management plan, and institutional support. The rest – 75 per cent of the total – proved to be nothing more than "paper parks". In most cases their borders are not known or respected by local populations, so the parks continue to be exploited.

Lack of coastal management in the Caribbean could begin to translate directly into lost tourist dollars. Already, studies reveal that nearly all of the coral reefs of the Caribbean have been damaged or disturbed to a certain degree. The only pristine reefs left are off the coasts of Belize and Panama, and around some of the more remote islands of the Lesser Antilles. If management plans do not begin to take hold during the course of the next decade, those tourists seeking more than a beach chair in the sun will take their holidays somewhere else.

When countries finally do draw up coastal management plans and carry out resource inventories, little of the work ever gets translated into action. Although the Environmental Management Bureau of the Philippines has prepared voluminous documents on the country's vast coastal zone and drawn up recommendations for its management, no action has been taken to implement the relevant recommendations.

Even projects which do include coastal management can be irritatingly inept in making the proper links with land-based problems which affect coastal areas. On the long, thin and largely undeveloped island of Palawan in the Philippines, an Integrated Management Project focuses on upland logging, with a coastal zone component. Unfortunately, the project has not developed an integrated approach towards resolving coastal development and related management problems. Both logging and tourist development continue, but there is little understanding of their impact on each other. At Bacuit Bay, the site of two tourist complexes, logging in the watershed has increased erosion by as much as 200 per cent; five per cent of the reef is already dead. Gregor Hodgson, a biologist working in the Bay, calculates that if logging continues over the next ten years, there could be a loss of $41 million in potential fisheries and tourism revenues.

Even Thailand, which has taken a leading role in resource management in Southeast Asia, has not been able to evolve working coastal-zone management strategies. The Thais have introduced coastal-zone development planning, but such plans contain little in the way of sustainable management of coastal resources.

Coastal-zone management may be difficult, but it is not impossible. There are successful management programmes in place that can be used as models. Probably the most acclaimed and complicated management plan ever devised and implemented is the one that governs Australia's Great Barrier Reef Marine Park. The Great Barrier Reef is by far the largest marine park in the world, covering

345,000 sq km – bigger than the United Kingdom. It extends for over 2,000 m along Australia's north-east coast, and includes 2,900 separate reef formations and coral islands, with 70 vegetated cays. The area also includes nearly 600 "high islands" located close to the mainland, which are managed by the State of Queensland. Technically, the region is not a park, but rather a multiple-use management system.

The Great Barrier Reef Marine Park was formed in 1975 by an Act of Parliament which had the support of all political parties. Public outrage at the prospect of offshore oil drilling and mining on the reef was the driving force behind the formation of the park. The Great Barrier Reef Marine Park Authority was established to administer the park, with broad powers to regulate and prohibit activities within its borders. Perhaps more importantly, the Act gave the Commonwealth Government the right to interfere in land-based activities that might threaten the integrity of the region's reefs.

But declaring the entire area a marine park was only the first step. Managing it in a sustainable way required comprehensive planning. Before a management strategy could be worked out, a complete resource inventory was taken and highly detailed maps of the entire reef system were drawn up. At the same time the Marine Park Authority sent out thousands of detailed questionnaires to individuals and organizations that used the reef regularly, in an effort to learn how the reef was used, where and by whom. The Authority followed up this process with public meetings and information campaigns, designed to enlist the support of the public and special interests (e.g.) recreational and commercial fishermen, and divers) for their proposed zoning plans.

Once the information was available, the Authority began the task of dividing the entire reef into management zones. In order to facilitate the management of such a vast area of sea and coast, the Barrier Reef itself was divided up into four huge management sections, running from north to south. Within each section, the Authority designated five different kinds of management zones, each prohibiting or permitting certain kinds of activities. These zones include:

- *General Use Zones*. These are all-purpose zones in which most activities, other than mining, drilling and bottom trawling, are permitted.
- *Marine National Park A Zone*. This is a kind of recreational zone where a variety of activities are permitted, including

general recreation, trolling for pelagic species, line fishing, and research;
- *Marine National Park B Zone.* This is a "look but don't touch" zone. All fishing is prohibited so that people may appreciate and enjoy the area in a relatively undisturbed state;
- *Scientific Research Zone.* As the name implies, this zone is limited to scientific research; and
- *Preservation Zone.* In these zones, no activities are allowed, with the exception of special scientific research which cannot be carried out anywhere else.

"The only activities that are not permitted in any part of the Barrier Reef are oil exploration, mining, littering, spear-fishing with SCUBA gear, and the taking of large specimens of certain species of fish," explains Graeme Kelleher, Chairman of the Great Barrier Reef Marine Park Authority, a post he has held since 1979.

Most of the sections have large areas of "General Use" zoning, where multiple activities are permitted. Still, the success of the entire management concept rests on the voluntary compliance of the general public and specialized user groups. "We cannot possibly patrol the entire reef with only 60 field staff and a few airplanes," points out Kelleher. "Some reefs are only visited once a year, despite the fact that we spend some $700,000 a year on aerial surveillance."

With limited staff and funding, the Marine Park Authority directs much of its efforts at educating the public about the different zones and the need for self-policing. So far, this tactic seems to be working. There are few violations of the zoning restrictions: only about 60–70 serious violations occur each year.

The Barrier Reef's main source of income is from tourism, which is developing rapidly. In 1986–7, for example, 160,000 tourists visited the region's 24 island resorts. By the year 2000, the Reef may see half a million visitors annually. Despite early fears that the Reef would be over-run with trophy-hunting tourists, so far the number of off-shore resorts has been kept in check.

Until recently commercial fishermen have been the park's "biggest human problem," according to Kelleher; "They opposed all restrictions and wanted the entire reef open to trawling." Fortunately, however, Kelleher has witnessed a recent improvement in co-operation from the fishermen and their organizations.

But by far the most controversial issue on the reef is the crown-of-thorns starfish, a natural predator with a taste for the coral polyps which build reefs. Normally, a few starfish feeding on coral polyps make little difference to the reef's ecosystem. However, the crown-of-thorns starfish has a disturbing habit of descending on reefs by the millions. Since each individual starfish can eat its own area of coral in a day, a large infestation can soon destroy an entire reef. Worse, each female crown-of-thorns starfish is capable of laying 20–100 million eggs. The offspring of just one individual, if they survive, are enough to cause a dangerous infestation. So far, nearly one-third of the entire Barrier Reef has been affected by this menace.

The problem is complicated by the fact that the crown-of-thorns has few enemies. Nothing seems to eat it, except the rare triton mollusc. So far, large-scale controls have been impossible to apply. Divers have been conscripted to inject starfish with poison and in some areas fences have been erected to keep them out. But given the scale of the invasion, these steps remain inadequate.

So far, there is no method of control which works, despite the $1 million being spent on research each year. "We have 30 research projects involving 50 scientists," claims Kelleher. "But we just don't know enough about this starfish to know whether or not general controls should be instituted. Any reef-wide control strategies that might work, would cost tens of millions of dollars."

Kelleher believes that the most serious problem facing the Great Barrier Reef in this decade will be the need to reduce run-off of sediments and nutrients from the mainland into reef waters.

Barring natural catastrophies, the multi-use management concept for the Great Barrier Reef seems to be a promising way to preserve coastal resources from over-exploitation, without declaring the entire area off-limits to any kind of use. The lessons learned by the Marine Park Authority in managing such a vast region could well be applied elsewhere.

Organizations like the International Centre for Living Aquatic Resources Management (ICLARM) in Manila seem to have learnt from the Great Barrier Reef Authority. But the coastal problems afflicting most Third World countries are far more complicated than those confronting the Great Barrier Reef. None the less, motivating local people to accept management plans is fundamental to the success of any coastal management strategy, an issue that ICLARM and other NGOs are beginning to address.

International NGOs, such as the International Union for Conservation of Nature and Natural Resources (IUCN), are helping developing countries draw up management plans for marine parks and protected areas. Presently the IUCN is active in 22 countries.

In some cases, as on the island of Negros (the Philippines), local communities are taking measures into their own hands without government or other outside help, to conserve and manage their coastal resources.

Unfortunately, much of what passes for coastal zone management misses the mark. Far too often, coastal management plans are not integrated with overall development objectives. There are few success stories, and nearly all of the sustainable projects tend to be local and small-scale.

Turning the tide of destruction that has beset coastal environments in every one of the regional seas that wash Third World coasts requires nothing less than a full-scale national programme of action, tied into regional-scale strategies for sustainable development of the sea and its coasts. In this sense UNEP's Regional Seas programme offers one of the best conduits for international, and above all regional, co-operation for combating common problems.

The threat of rising seas

One global problem that is receiving a great deal of attention from the United Nations, international NGOs, governments, academics, the media, even local citizens' groups, is the imminent threat of climate change and sea level rise. If this obsession with the likely impacts of sea level rise – especially for low-lying islands and coral atolls – generates better management of coastal areas, so much the better. It is the only coastal issue, for the moment, which is capable of arousing a truly international response. In order to address this threat in a realistic manner, coastal nations will also have to begin to manage coastal resources and near-shore waters. Leaving things to chance, or trying half-measures, will not do; if the sea level is going to rise, it will. Some scientists, like Dr Irving Mintzer, Director of Policy Research at the Center for Global Change, University of Maryland, believe we are destined for a temperature increase and expanding sea levels, no matter what we do. The questions are: by how much, and where will the worst affected areas be?

Greenhouse gases – carbon dioxide, methane, nitrous oxide tropospheric ozone and chlorofluorocarbons (CFCs) – already released into the atmosphere insure a temperature rise of anywhere from 2–5° Fahrenheit by early next century. "This may not sound like much of an increase," observes Mintzer. "But a warming of just two degrees would take the planet outside the temperature range which has been experienced in the last 10,000 years." The seas have already risen by 20 centimetres, due to the thermal expansion of seawater when heated. More incremental rises are expected as temperatures increase and icecaps begin to melt as well.

UNEP, along with a host of other national and international organizations, are busy compiling reports and studies that attempt to answer the questions posed by climate change. There is a sense of urgency. Even a rise of as little as one metre would inundate coastal lands, particularly river deltas, mangrove forests, and coral islands. A billion people could be turned into environmental refugees by a 1–2 m rise in the world's seas. According to Stjepan Keckes, Director of UNEP's Oceans and Coastal Areas Programme Activity Centre in Nairobi, "sea level rise for many island states and territories, as well as some mainland nations, could be the most serious environmental problem confronting them over the course of the next century."

Speaking at the Commonwealth meeting in Kuala Lumpur, Malaysia in the autumn of 1989, the President of the Maldives, Mr Maumoon Abdul Gayoom, told the assembled heads of state that if the sea level rose as predicted by one m, by 2050, many countries in the Commonwealth "will simply disappear". With most of his nation's islands no more than two m above sea level, the Maldives is, in Gayoom's words, "an endangered country".

Similarly, Bangladesh reports that a one m rise in the Bengal Sea would flood 15 per cent of the entire country, killing off thousands of hectares of mangrove forests in the Sundarbans and displacing at least 10 million people. Tens of millions would be forced to flee the floodplains of the Indus River Delta in Pakistan and the Nile Delta in Egypt.

A UNEP report detailing the effects of sea level rise in the South Pacific claims that those of the region's 5 million people living on coral islands and atolls could suffer drastic consequences from a 1–2 m rise in the Pacific. Many island countries, such as Tuvalu, Kiribati and the Marshall Islands, would be swamped or under water, their populations forced to move to higher, drier islands or the mainland countries around the Pacific rim.

Rising seas could flood over a quarter of Papua New Guinea's 17,000 km coastline, inundating floodplains and river deltas. A one m rise would lead to saltwater intrusions into water tables, the waterlogging of soil and massive coastal erosion.

With rising seas come increased effects from storm surges, like hurricanes and typhoons. A one m rise in the world's seas could mean terrible consequences for mainland coastal populations, like those of Bangladesh and India. Storm surges in the late 1970s killed 300,000 Bangladeshis and reached 150 km inland. A higher sea level would translate directly into higher death tolls and escalating property damage from storms.

The same could be said for virtually every one of the smaller and lower-lying islands of the Caribbean Sea and Indian and Pacific Oceans. It took a full half year before Jamaica recovered from Hurricane Gilbert, which struck the island in 1988. A hurricane which swept over Dominica in 1979 destroyed 85 per cent of the island's houses; a typhoon in Tonga in 1982 wiped out half the island's houses and devastated that year's fruit crop. Imagine how much worse these storms would have been if the sea had been a metre higher.

With just such a scenario in mind, Mr Gayoom of the Maldives proposed to the Commonwealth meeting in Kuala Lumpur that since the poorer countries of the developing world had nothing to do with causing this human-induced catastrophe, the rich nations should mobilize international assistance for those countries most vulnerable to climate change and rising seas. Something will have to be done, warned Mr Gayoom, unless the world wants to be "inundated with environmental refugees", most of them poor.

The Prime Minister of Malaysia, Mr Mahathir bin Mohamad, even proposed that an international fund be established to help developing countries pay for imported technologies, reforestation schemes, and other programmes to protect the environment from the effects of global warming and inexorably rising seas.

One of the best results that could emerge from all the concern over climate change and swollen seas would be the development of workable coastal management plans for all developing countries with a coastline. Proper, well-conceived and executed coastal zone management strategies would permit billions of people to utilize their resources better and give them time to prepare for rising seas. Such plans would also allow coastal populations to build for a sustainable future. Perhaps more importantly, they might help give millions of the poorest people, those constantly living on the edge of survival, a sustainable present.

2. The Regional Seas of the Developing World

When the United Nations Environment Programme (UNEP) began the first of its regional seas programmes, in the Mediterranean Sea in 1974, few officials then could have predicted its evolution into one of the most comprehensive UN programmes ever launched to control marine pollution and promote better management of coastal and near-shore resources. After its auspicious beginning in the Mediterranean, the programme went on to encompass 10 of the world's regional seas, involving 120 countries. With the sole exception of the Mediterranean coast, which includes both highly developed and barely developing economies, all of UNEP's efforts at setting up regional mechanisms for managing seas and coasts in a sustainable manner have been concentrated in the South. The following chapters describe each of these seas in turn.

Despite initial successes in setting up regional programmes – the Mediterranean, Persian Gulf and the Caribbean, for example – the record is a very uneven one. The beginning promised much. Many governments were quick to recognize the value of a regional seas approach in confronting the crisis of their coasts. But following up lofty proclamations with skilled labour, money and, most important of all, management plans, took years of painstaking work by UNEP and the relevant states. Many countries involved in the regional seas network have yet to evolve coastal management strategies or put them in place. Of the ten regional seas programmes currently operating, perhaps four – the Mediterranean, Wider Caribbean, the Gulf of Arabia and the South-east Pacific (west coast of South America) – can point to some progress in setting up sustainable management regimes. None of the programmes, with the notable exception of the Mediterranean, has had much success in combating pollution. But even the Mediterranean states trod water on this issue; it was a decade before anti-pollution measures began to take hold.

This is not to say, however, that the other programmes have run aground. In many regions, with their nationalist rivalries and history of confrontation, the simple fact that a regional forum exists for dealing with common environmental concerns is something to celebrate. South Asia, for instance, has not yet endorsed an action plan to protect and manage its seas and coasts. Yet the first hurdles have been overcome. Structures have started to develop for dealing with the problems on regional level. More importantly, perhaps, the very fact of launching regional seas programmes has strengthened national institutions charged with managing coastal and sea concerns. At the least, research capabilities have improved and the training of specialists has been given some priority.

Institution building takes time, especially for poor countries with no planning or management foundation to build on. Dr Ed Gomez, Director of the Marine Science Institute at the University of the Philippines in Manila, believes that scientific research into Southeast Asia's marine and coastal problems has been greatly enhanced because of the regional seas programme. "We would certainly not be this far along in our research without COBSEA, the regional seas programme for this region." The same could be said for virtually every one of the other programmes, no matter how far along they are in tackling their common problems. In many regions, there is no network of scientific institutions dealing with marine research and related problems. Scientists have had to form informal contacts with colleagues in neighbouring countries and elsewhere. At least the regional seas network provides a forum for scientists and policy-makers to exchange information and ideas.

Peter Thacher, former Deputy Executive Director of UNEP, describes the underlying factors which influence UNEP's marine policy:

> There are two fundamental notions in back of the regional seas philosophy. The first is the idea that while marine pollution problems are global in scale, the most efficient way to solve them might be through co-operative programmes at the regional level. The second is that although marine pollution problems are perceived to be somewhere "out there", off-shore and over the horizon, the origin of most of them is in fact here on dry, sovereign land.

By the time the Stockholm Conference on the Human Environment took shape in 1972, there was already a consensus among policy makers and scientists that land-based activities were a significant source of marine pollution. But getting the issue of ocean pollution firmly on the Stockholm agenda was largely the work of Maurice Strong, UNEP's first Executive Director and one of the movers behind the environment movement. In preparations for the Stockholm Conference, Strong recommended the formation of a working group on marine pollution for the following reasons:

> First to establish the broad design for a master plan to safeguard the health of the oceans for the greater benefit of all mankind, and second to initiate action with a view to intergovernmental agreement by the time of the Stockholm Conference on some particular measures which are both urgent and feasible, such as a convention on the control of ocean dumping.

After UNEP was established, as a result of the Stockholm Conference, one of the first mandates given to the fledgling organization was to come up with a programme of action to reduce marine pollution and loss of vital coastal resources.

At the same time, the Stockholm Conference took the issue of the health of the oceans even further and endorsed, as Strong had suggested, what came to be known as the London Ocean Dumping Convention, a legally binding treaty to curtail the dumping of dangerous and hazardous substances in the world's seas. It was opened for signature in early 1973, less than a year after the Stockholm Conference.

Naturally, the momentum of Stockholm gave impetus to initial efforts at setting up the first regional seas programme in the Mediterranean. It was the obvious sea with which to open the programme. Despite its many languages and different cultures, most Mediterranean peoples think of the sea as *mare nostrum*. Peter Thacher recalls that representatives from a number of Mediterranean states, who would not talk to each other, nevertheless approached UNEP to urge that something be done. "At the time there were outright hostilities between various states from one end of the Mediterranean Basin to the other. This made dramatic the hope that if – a big 'if' in those days – agreement could be won to allow governments to co-operate in preserving their common future, it might have beneficial effects on current, day-to-day relations between them."

The framework established for the Mediterranean Action Plan has been copied elsewhere. Although each programme reflects different priorities from region to region, they all share some common ground. Thacher stresses that "each has, for example, a common table of contents setting forth co-operative work in scientific, economic, political/legal areas with a section on the institutional and financial means by which to ensure that action is taken."

But progress in implementing the action plans varies a great deal. It may have taken only one "power lunch" for Dr Tolba, the current Executive Director of UNEP, to pull together a financial arrangement for the Persian Gulf, but others, like the Caribbean, were stymied for years behind financial constraints. Many of the programmes are still hampered by slack budgets and institutional inertia.

Too many governments have yet fully to recognize the crisis of their coasts. A kind of resigned lethargy has settled over some of the regional seas programmes, as national institutions dealing with seas and coasts struggle to get their priority issues onto the political agenda, with only limited success. In many cases, agencies or ministries charged with environmental concerns do not even have direct access to the levers of power. They are isolated and removed from the give and take of governance. Environment is the unwanted step-child of development.

Until the crises which threaten to overwhelm sea coasts are given credence, progress towards confronting the challenges will be desultory. In some regions they may peter out altogether as other more pressing economic matters consume budgets and staff resources. Meanwhile, the regional seas programmes, along with other national and international efforts to resolve the problems caused by growing populations and loss of coastal resources, will continue to sound the alarms. The question is: will they be heard in time?

3. The Mediterranean Sea

The Mediterranean Sea is the centre of western civilization. Along its shores flourished some of the world's greatest empires: Egyptian, Persian, Phoenician, Assyrian, Minoan, Macedonian, Greek and Roman. It was (and still is) a cauldron of cultures and has always been a vast highway over which goods were traded and people transported. It was a vital link in the spread of cultures, political systems and religions throughout the world. Even 100 years ago, Homer's wine dark sea seemed vast and unconquerable.

The modern Mediterranean, however, has shrunk. It has become, in the words of French historian Paul-Marc Henry, "a small, crowded lake, and a polluted one at that". By the mid-1970s the Mediterranean had been turned into a huge waste-bin, a receptacle for millions of tonnes of pollution. Its shores were scarred with unbridled development; its cities and towns bursting at the seams with people. At the same time, the sea was being transformed into one of the world's favourite summer playgrounds. During the summer months upwards of 100 million tourists scramble for its beaches.

Recognizing the urgent need for regional action to save the Mediterranean, UNEP launched its first of ten regional seas programmes. In 1975, the Mediterranean states agreed to an Action Plan for the Protection of the Mediterranean Environment. The following year, delegates from all 18 Mediterranean countries (except Albania) signed the Barcelona Convention for the Protection of the Mediterranean Sea Against Pollution, plus two protocols. It was the beginning of a long and complicated process to reverse the degradation of the sea.

For UNEP, the Mediterranean Action Plan (MAP) was a diplomatic coup. Within the context of the plan Greek, Turkish and Cypriot marine biologists have agreed to see their coasts as an environmental issue, rather than a political one. Israeli sanitary engineers have sat down with their Arab counterparts. The plan remains the only international forum through which regional co-operation in the Mediterranean Basin is made possible.

Too much geography, too little sea

In a sense, the Mediterranean Sea is a victim of unfortunate geography. Although it is relatively deep (averaging 1,500 m), it is not a very big sea, covering an area of about 2.9 million sq km. The volume of water amounts to 3.7 million cubic km. Impressive as these numbers sound, they don't add up to much. The Mediterranean is too far north to support coral reefs and mangrove swamps. And it is not blessed with vast submarine meadows. Continental shelves are non-existent; shallow coastal waters suddenly plunge to depths of 1,000 m or more. Instead, the sea is nutrient-starved and relatively fish-poor.

The lands around the Mediterranean are as barren as the sea. Low rainfall in many areas means agricultural land must be irrigated. Except for Libyan oil, it is poor in mineral and energy resources. In the dry eastern and southern parts of the Mediterranean Basin, water is often more expensive than oil.

Because of its unfortunate geography, pollution tends to accumulate in the Mediterranean Sea. As any map shows, it is a confined, enclosed sea, bottled up behind the Strait of Gibraltar. Its hydrological pump works slowly, taking at least 80 years for its waters to renew themselves. Also, since there are no tides to speak of, there are no tidal currents to flush away pollution.

Crammed along the 45,000 km of Mediterranean coastline is a permanent population of around 130 million, supplemented every year by over 100 million tourists. Demographers predict that its resident population may soar to 200 million by the turn of the century.

Most of the region's big cities are located on or near the coast. In order to cope with the tourist boom, which stretches from May to October, entire coastlines have been purged of their natural habitats and reshaped to accomodate the annual invasion of northern tourists. For nearly half the year the population of the region doubles. The result has been widespread devastation of coastal resources, particularly wetlands, forests and the Mediterranean Maquis (a shrub-dominated ecosystem somewhere between grasslands and forests). But croplands have also been lost to urban expansion and the helter-skelter development of resort hotels and other tourist accommodations. In Spain, France, Italy and parts of Greece there are practically no undeveloped coastlines left. Along the Catalonian coast of Spain, only a small fraction of its 580 km-long shore has been spared from urban, port, industrial or tourist development. Seventy-five per cent of Italy's Romagna coast is "developed". Twenty years ago the south coast of Attica between Athens and Cape Sounion was nothing but olive groves and croplands. Today it is dominated by inter-connected tourist complexes and summer villas. Unpaved roads running back from the coast into scorched hills are lined with garbage, construction rubble and household junk.

Development along the Mediterranean has been lopsided. The booming economies of the industrialized northern rim countries stand in stark contrast to the poor, developing economies of the southern rim. By the mid-1970s most of the northern rim countries were highly industrialized, while Egypt and Algeria were just beginning to enter the industrial age. The populations of the southern rim countries – Egypt, Algeria, Libya, Morocco and Tunisia – are growing by two per cent a year and above, while the populations of the North are stabilizing, or growing at almost imperceptible rates (with the exception of Turkey).

There is also a religious divide between the Christian north- and Muslim south-rim countries.

Coastal pollution

Since the 1960s, the Mediterranean has acquired the reputation of being one of the most polluted seas in the world. Millions of tonnes of pollutants, both natural and synthetic, are spilled into the sea every year, mostly via river systems. Only some 30 per cent of the sewage from over 700 coastal towns and cities receives any form of treatment before being discharged into the sea. Often sewage outfalls don't extend far enough from shore, their contents spilled in shallow, near-shore waters where they create health hazards. The famous marine explorer Jacques Cousteau once described the sea as a veritable "garbage dump".

World Health Organization (WHO) scientist Richard Helmer breaks down the annual input of some (but not all) pollutants entering the Mediterranean from land-based sources as follows:

- 12 million tonnes of organic matter
- 320,000 tonnes of phosphorus
- 800,000 tonnes of nitrogen
- 100 tonnes of mercury
- 3,800 tonnes of lead
- 2,400 tonnes of chromium
- 21,000 tonnes of zinc
- 12,000 tonnes of phenols
- 60,000 tonnes of detergents
- 90 tonnes of organochlorine pesticides.

Most of this pollution is transported into coastal waters by the region's 70 major river systems (and numerous small streams). Not only does the coastal population of the Mediterranean foul its own nest, but much of the municipal and industrial wastes of the entire basin find their way into the sea as well.

Heavy metals such as mercury, lead, zinc, chromium, copper and cadmium continue to enter the Mediterranean in unregulated quantities. Since toxic metals are all considered persistent poisons, accumulating in food chains and bottom sediments, they represent chronic health threats in certain areas. Filter-feeders like clams, mussels, and oysters are known to concentrate heavy metals in their tissues. High levels of copper have been found in oysters in the Rio Tinto Estuary in Spain, and mercury concentrations of up to 47 ppm (parts per million) have turned up in bottom sediments along the southern coast of Tuscany (Italy) and in the northern part of the Gulf of Trieste. High mercury levels have also been found in commercially important fish and shellfish such as bluefin tuna, striped mullet and Norway lobster. Most of these metals are brought by rivers, but some fall from the atmosphere attached to minute particles of soot and dust.

In general, the developed northern rim is more contaminated with pollution than the less-developed southern rim. However, nearly every one of the Mediterranean's 70 major cities (those having more than 100,000 people) create local "hot spots" of pollution from the discharge of untreated sewage and industrial wastes. Chronically high pollution levels are recorded in the northern Adriatic, Izmir Bay (Turkey), Elevsis Bay (Greece), in the Lagoon of Tunis, and in the coastal waters of Alexandria, Egypt.

Elevsis Bay, outside Athens, is said to be suffocating from industrial pollution pumped into its waters from 29 industrial enterprises along the northern coast of Attica. Untreated effluents from shipyards, refineries, steel mills, electrochemicals, canneries, textiles, and glass and cement works have rendered the Bay's once clear waters a dingy brown. Most marine life has been exterminated.

The coastal waters of Alexandria are in worse shape. Around one million cubic metres of untreated municipal and industrial wastes are discharged every day through outfalls too short to allow these wastes to be diluted or dispersed. Instead, the entire waterfront of Alexandria reeks with rotting sewage, algal scum, household slops, construction debris and industrial wastes. By the summer of 1986 bathers along Alexandria's beaches began complaining of pollution-related skin sores, rashes, and intestinal disorders.

A $500 million joint US–Egyptian project, called the Updated Alexandria Wastewater Master Plan, aims to renovate and upgrade two existing sewage treatment plants, neither of which functions at present. But even when these plants start operating sometime in 1991, some 300,000 cu m of sewage a year will still escape treatment. A UNEP task-force was sent to the city to advise municipal authorities on how to treat and dispose of their growing mountain of wastes. So far, no action has been taken.

Despite clean-up efforts, the Mediterranean is still shamefully polluted. Sea-side vacationers risk everything from skin rashes to viral hepatitis and food poisoning (salmonella) from bathing in dirty waters and eating contaminated fish and shellfish. One-fifth of all beaches are still so polluted as to be unsafe for bathing. Athenians are now accustomed to driving 70 kilometres to find clean beaches. And many of the region's shellfish growing areas do not produce seafood fit for the table. Typhoid is said to be 100 times more frequent in the Mediterranean than in northern Europe. Even periodic outbreaks of cholera have been reported over the years. In the summer of 1973 a cholera epidemic struck Naples and southern Italy, with 325 reported cases and 25 deaths. The cause was traced to mussels contaminated with raw sewage.

The other black death

The scourge of oil pollution, on the other hand, knows no boundaries. Because the Mediterranean is one of the major crossroads for the world's merchant and tanker fleets, oil is found nearly everywhere. Both north and south suffer. Until recently, the Mediterranean Sea was regarded as one of the most oil-fouled bodies of seawater in the world. Philippe Le Lourd, former Director of the Regional Oil Combating Centre in Malta, estimates that somewhere between 500,000 and one million tons of oil (and oil-related products) are flushed into the Mediterranean every year. Most of this pollution comes from routine shipping operations and tanker traffic through the discharge of dirty ballast waters, bilge slops, and oily wastes.

In the mid-1970s, the Mediterranean was covered with so much oil that scientists trying to study marine life often pulled in nets filled with tar balls instead of fish. In 1977, during a research cruise in the eastern Mediterranean, scientists from the International Atomic Energy Agency's laboratory at Monaco made the following observation: "between Crete and Libya a 30-minute neuston (surface) tow completely filled a one-litre collecting jar with tar balls. Oil and tar residues were omnipresent and great caution was necessary when lowering plankton nets through the surface film to avoid contamination." In other areas, as much as 500 litres of tar per sq km of water surface have been found. The largest concentration of tar was found in the Ionian Sea off Libya and between Libya and Sicily. By 1980, the Mediterranean Sea accounted for nearly a fifth of all oil spilled into the world's oceans from accidents and routine shipping.

Today, around 600 million tonnes of petroleum products are shipped into and through the Mediterranean every year. About half of this amount ends up at 18 ports scattered around the sea. Although still a general problem, oil pollution has been reduced thanks to the installation of oil-waste handling facilities at major ports around the region.

De-oiling the Mediterranean

Although a Regional Oil Combating Centre was established on Malta, to co-ordinate efforts at fighting oil pollution from accidents, this strategy left unsettled the matter of routine oil pollution from normal shipping activities. Action on this front got underway in the early 1980s, led by Greece, whose maritime tradition goes back 3,000 years.

Instead of opting for land-based facilities, the Greek government decided to promote the construction of floating facilities for the reclamation of waste oils and other petroleum products. Today, the country has seven such facilities scattered around the country at major ports like Piraeus, Patras, Siros, and Thessaloniki.

Both the International Maritime Organization (IMO) and the European Community support the construction of floating facilities. Given the success of the Greek operations – in 1987 Greece was able to collect nearly 2.6 million tonnes of oily wastes, from which they extracted over 26,000 tonnes of recovered oil – more of them are expected to be set up in ports around the Mediterranean and elsewhere. Recently, two more went into operation: at Rejeka in Yugoslavia and at Port Said in Egypt, where 20,000 ships pass on their way through the Suez Canal every year.

The first oil reclamation facility in Greece was built by a young entrepreneur named Denis Yatras, using, appropriately, a retired oil tanker. His ship, the *Delta*, is currently anchored in Elevsis Bay wedged between huge ocean-going tankers. Yatras explains how his operation works. "In 1987 we processed over 400,000 tonnes of tanker slops and bilge water and were able to get 17,274 tonnes of recovered oil. We then deliver our oil, which still contains three per cent water, to a local refinery, where it is re-processed. In return for the oil, we receive 75 per cent of the value of the oil in the form of fuel oil, which we market ourselves." This swap system seems to work well. After all, Yatras does not need to pay for the oily wastes, only for the maintenance of his ship and oil-water separator system, and the salaries of his engineers. So far, the *Delta* pays for itself.

Meanwhile, Yatras is working on a more sophisticated computer model, which will enable him to squeeze more recoverable petroleum products out of oily wastes from shipping. His goal is to design a system that will get better separation of oil from water, using less energy than his current process. He intends to market the system worldwide.

Yatras is convinced that floating oil recovery systems are the wave of the future. "The point is not just recovering oil, but pollution prevention. Just taking the 17,000 tonnes of oil that the *Delta* recovered in 1987 means that 1,700 square kilometres of sea are not polluted with oily residues one centimetre deep." According to the IMO, all major ports are now required to have oil-waste handling facilities. Unfortunately, requirements are often one thing, reality another. Too many ports have yet to introduce even basic oil-water separators.

Fisheries

The Mediterranean is a fish-poor sea. The 5.5 million tonnes of fish caught every year amount to a mere two per cent of the world take. There is no real fishing industry in the sea. Ninety per cent of all fishing activity is labour-intensive and small-scale. Most fishing boats are no longer than 30 feet, with crews of 2–4 people. Net fishing is concentrated in shallower, near-shore waters and around the sea's many islands.

Nevertheless, over the past 20 years a deep sea trawling industry has evolved. Concentrated around Sicily, these vessels can stay at sea for weeks and harvest great quantities of pelagic fish. The trouble is that over half of what they catch is tossed overboard because only certain kinds of fish, like tuna, fetch the high prices necessary to pay the overhead on large trawlers. Today, deep sea fishing around the Mediterranean is becoming more difficult as more and bigger ships are needed to bring home the same catches.

Fishing pressures are becoming intense along much of the northern rim, and along the coasts of Algeria, Morocco and Egypt as well. In many areas catches of popular shellfish, such as lobsters, have declined due to over-exploitation.

Because the Mediterranean's meagre fish catches cannot keep pace with demand, nearly two-thirds of all fish consumed in the region are imported from other seas, most notably the Atlantic. Still, the price of Mediterranean fish remains high. Its limited fisheries are worth about $700 million a year – a full five per cent of the value of the total world catch of fish and shellfish. As long as fish stocks remain low and hard to get, the artisanal nature of Mediterranean fisheries will prevail. Increasingly, big trawler operations cannot stay out long enough to take in profitable catches.

Marine mammals

Like marine mammals everywhere, those in the Mediterranean are under siege. Dolphin populations have fallen, despite conservation efforts, as have the numbers of sea turtles and seals. In most cases, pollution is not the problem. Marine mammals are running out of space as their breeding habitats are destroyed or disturbed to make room for more people.

Perhaps the most critically endangered marine mammal in the region is the Mediterranean monk seal (*Monachus monachus*). With a current population hovering around 600, it is restricted to a few isolated Greek islands and the Aegean coast of Turkey in the eastern Mediterranean. Unfortunately, the monk seal is a victim of both habitat loss and commercial fishing pressures. Fishermen view the seal as a competitor for limited fish stocks and kill them whenever they can. Others are strangled and drowned when they become entangled in fishing nets. Although there may be more conservation groups trying to save this animal than there are seals to save, the few remaining monk seals stand little chance of survival, unless local fishermen are enlisted for their protection.

Land degradation

In many areas of the Mediterranean Basin, land is being degraded from over-cultivation, deforestation, poorly planned and installed irrigation systems, and the spread of deserts.

The arid zone, which incorporates the south-eastern part of Spain, Morocco, Algeria, Tunisia, Libya, Egypt, Israel and Syria, suffers principally from soil erosion and advancing deserts. In southern Tunisia alone an area of 12,500 sq km was lost to the Sahara in less than ten years. The reasons were attributed to inappropriate agricultural practices, over-grazing of rangelands by domestic animals, and poorly managed irrigation schemes. In this area, the loss of topsoil averaged ten tonnes per hectare per month. On the fringes of the Sahara Desert, soil losses as high as 250 tonnes per hectare per year have been recorded.

Deforestation consumes around 60 million hectares of shrub and forestland every year. Much of this wood, perhaps as much as 60 million tonnes, is burned for fuel. Although some of this degraded forestland regenerates itself during rainy years, vast tracts remain treeless, and increasingly desertified. As more land becomes degraded, more people are forced off the land and into coastal areas.

In addition, somewhere between 30,000 and 40,000 hectares of irrigated croplands are lost every year due to salinization and alkalization of the soil. These processes effectively sterilize the land, rendering it unfit for agricultural production. Degraded land can be reclaimed, but it is expensive and time consuming.

The Mediterranean Action Plan

In 1975, 16 of the Mediterranean's 18 states gathered in Barcelona for an historic occasion: the approval of an Action Plan for the Protection of the Mediterranean Environment. The plan called for a broad three-pronged approach to controlling pollution and managing the sea's resources collectively: a series of legally-binding treaties to be drawn up and signed by Mediterranean governments; the creation of a pollution monitoring network and the co-ordination of scientific research efforts; and a socio-economic programme that would reconcile vital development priorities with a healthy Mediterranean environment.

The following year, those same countries, plus the European Community and Syria, returned to Barcelona in February to sign the Barcelona Convention for the Protection of the Mediterranean Sea Against Pollution. The convention committed the Mediterranean states to "take all appropriate measures . . . to prevent, abate and combat pollution . . . and to protect the marine environment". It provided the legal muscle for the action plan which had been adopted in 1975. Two significant protocols were also signed on 18 February 1976. The "Protocol for the Prevention of Pollution of the Mediterranean Sea by Dumping from Ships and Aircraft" blacklisted dangerous wastes such as mercury, cadmium, crude oil, chlorinated hydrocarbons, pesticides and radioactive substances, prohibiting their disposal in the sea. A grey list of substances that could be dumped under special circumstances (by permit only) was also included. A second "Protocol concerning Cooperation in Combating Pollution of the Mediterranean Sea by Oil and Other Harmful Substances in Cases of Emergency" commits governments to co-operate in combating oil and chemical spills. In 1976 a regional oil-combating centre was established in Malta as part of the Action Plan.

In many cases, the protocol on oil pollution was buttressed by national legislation already in place. Greece, for example, has an extensive set of laws protecting the marine environment from oil pollution. "Every port authority in Greece – some 90 in all – has plans and equipment for combating oil spills," states Captain D. Doumanis, Head of the Marine Environment Protection Division in the Ministry of Mercantile Marine. "This has helped enormously in lessening the coastal impacts of oil spills."

As the first step in regional co-operation to combat pollution, the participants in the Action Plan agreed to set up the Mediterranean Pollution Monitoring and Research Programme (called MEDPOL). The first phase, which ran from 1976 to 1980, involved scientists from 83 research centres and laboratories from 16 Mediterranean countries. The initial years were devoted to finding out just how polluted the sea really was by conducting numerous baseline studies. The Action Plan countries also established standardized analytical procedures so the data gathered by one institution, in say Egypt, could be utilized and understood by colleagues in France or Italy.

By introducing standardized pollution monitoring and data collection methods, a scientific network could be established throughout the region and expensive duplication of research avoided. By 1985, 102 research projects were being carried out by 62 research centres in 16 Mediterranean countries. Today, 14 countries have implemented national monitoring programmes in co-operation with UNEP.

In 1979 a "Blue Plan" for the long-term management of the Mediterranean Sea was launched as part of the socio-economic component of the Action Plan. It was intended to "take the long view", dovetailing future development plans with environment protection measures. A regional centre for co-ordinating Blue Plan research was set up in Sophia Antipolis, on the French Riviera.

The centre has issued numerous reports dealing with a wide spectrum of issues, including freshwater management, industrial growth and industrialization strategies, energy needs, population movements, urbanization, and the impact of tourism on coastal areas. One future scenario elaborated by Blue Plan research predicts that, if present trends hold, by the middle of the next century the Mediterranean might see 95 per cent of its coastline urbanized to meet the needs of a resident population that has swollen to 200 million, supplemented by 200 million seasonal tourists. Roads, by then, would have to accommodate 150 million private cars (compared to the current 50 million). Balancing the needs of such a staggering number of people with environmental considerations will require regional co-operation on a level not yet achieved.

In 1980, after lengthy negotiations, the 17 members of the Action Plan signed the Protocol for the Protection of the Mediterranean Sea Against Pollution from Land-Based Sources. This landmark agreement identifies measures to control coastal pollution from municipal sewage, industrial wastes and agricultural chemicals. It also has a black and grey list of substances similar to those noted in the dumping protocol.

Each government, however, is responsible for implementing the protocol in its own way. Good intentions are not enough. With the total costs of coastal cleanup estimated at $15 billion over the next 10–15 years, a number of the area's cash-short countries have yet to act substantively on this mandate.

At a meeting in Geneva in 1982, the Mediterranean governments also approved a protocol providing special protection for endangered species of fauna and flora as well as critical habitats. Officially called the Protocol Concerning Mediterranean Specially Protected Areas, it entered into force in the spring of 1987, after nine countries and the European Community had ratified it. By 1988 it had been ratified by 13 countries. A regional centre was established in Tunis to assist governments in applying the protocol.

The agreement binds signatories to set up areas of biological interest – fisheries, breeding grounds for protected species, and monk seal and sea turtle sanctuaries – within the next few years. Under the umbrella of the protocol, a number of marine parks and protected areas will also be created.

Through UNESCO's International Council on Monuments and Sites (ICOMOS) Mediterranean countries have identified over 100 special historic sites that merit protection. The sites, which have to be of common interest to the peoples of the Mediterranean – not just nationally significant – range from Malta's pre-historic megaliths, which predate Stonehenge by a thousand years, to the beautiful walled city of Rhodes. The sites were approved by all of the Mediterranean Action Plan countries, who have agreed to help preserve and protect them.

Action in the Action Plan

Despite periodic financial setbacks, the Mediterranean Action Plan (MAP) co-ordinating unit, headquartered in Athens, is currently operating with an annual budget of $5 million. Although the Action Plan functions as part of UNEP, programme decisions are taken by the governments of the region. Since MAP is now self-sufficient, UNEP no longer contributes funds for its operation. "Instead," explains Aldo Manos, director of MAP, "the programme now pays UNEP $500,000 a year for co-ordinating activities in Nairobi." The 17 countries of MAP set up a Mediterranean Trust Fund to underwrite costs for the secretariat in Athens, the five regional centres, and to provide core funding for the Plan's far-flung research, monitoring and conservation programmes.

"We now have strong political commitment on the part of our supporting governments," states Manos. "Countries also give additional contributions to fund specific projects of interest."

Although the first decade of MAP was noted mainly for its accumulation of data, the action part of the Action Plan is beginning to get results. MEDPOL has moved into the management of coastal resources. In addition, most of the countries of the region have begun to take concrete measures to reduce pollution from land-based sources under the terms of the protocol mentioned above.

At a meeting in Genoa in 1985, the MAP countries agreed to sweeping measures in a co-ordinated effort to make the Mediterranean Sea cleaner and safer. Ten priority targets were set for the second Mediterranean decade (1985–95). Included among them were commitments to establish more reception facilities for dirty ballast waters and oily wastes from shipping; the construction of sewage treatment plants in all cities with over 100,000 inhabitants, as well as the installation of outfalls and/or treatment plants for all towns with more than 10,000 residents; the application of environmental impact assessments in coastal development; co-operation to improve the safety of maritime navigation, particularly for those ships carrying dangerous or highly polluting cargoes; and measures to reduce industrial pollution of coastal waters, as well as the proper disposal of solid wastes.

Despite foot-dragging, a number of countries have begun to build sewage treatment plants for their large cities. Istanbul is spending $2 billion to clean up the once glorious Golden Horn, contaminated with raw sewage, slaughter-house offal and industrial wastes. Marseilles recently inaugurated a giant $180 million underground sewage treatment plant to handle the municipal wastes of the city's one million inhabitants. The European Investment Bank has allocated $5 billion to clean up the Po River in Italy, a major source of pollution of the northern Adriatic. And in Athens a huge sewage treatment plant on the island of Psitalia, which will process all municipal wastes from the city's 4 million inhabitants, is nearing completion.

Between 1983 and 1985 Spain spent 7 billion pesetas (about $40 million) on beach protection, access roads and pedestrian paths. But nearly five times that amount was expended between 1975 and 1985 on water supply, sanitation and sewage networks for 181 communities along Spain's Mediterranean coast.

In the summer of 1988, certain countries began to issue public reports on the state of their bathing beaches. Until then, the condition of most beaches was considered a state secret. No government wanted to disclose pollution data for fear that tourists would go elsewhere.

"All that has changed," asserts Manos. "France and Italy have taken the lead in being open and forthright about the state of their beaches. They have, so to speak, gone public. In France, for example, the government now publishes a booklet on the state of all its hundreds of beaches, describing their condition. It's also available in three languages." Other Mediterranean countries are expected to follow suit.

Still, most governments refuse to provide data on sourced points of pollution, such as factories or sewage outlets. "This remains a problem for us," confesses Manos, an Italian lawyer born on the Dalmatian coast of Yugoslavia. "Governments will not even give us hard data on pollution from specific rivers flowing into the Med. What we have are regional totals." Many countries still hide behind collective pollution figures; what one Mediterranean scientist calls "collective guilt".

Manos doubts whether the MAP countries will achieve all the goals of the Genoa declaration by 1995, but some will be accomplished. And there are, of course, numerous efforts by national non-governmental organizations (NGOs) designed to assist in the cleanup of the Mediterranean. Many small coastal towns in Italy, for example, no longer sell plastic bags with purchases, only paper ones. This is the result of local NGO action. Beach cleanups have been organized by citizen action groups. One of the biggest challenges facing MAP is how to harness the tremendous energy of environmental NGOs around the Mediterranean Basin and channel it where it will do the most good.

4. The Persian Gulf

Were it not for vast quantities of oil and gas lying around and under the Gulf, the countries of the region might still be poor and developing. Instead, oil wealth has projected them into the modern age. The eight Gulf States – Kuwait, Bahrain, Iran, Iraq, Oman, Qatar, Saudi Arabia, and the United Arab Emirates – sit above the largest reserves of oil in the world, some 51 billion tonnes of it. Ironically, in a region dominated by fierce political, economic and religious rivalries, oil is the only thing most of them have in common. The other common definition is desert, which stretches from North Africa across the Red Sea through the Arabian Peninsula to the Persian (or Arabian) Gulf and across to Iran.

Historically, the Gulf States are better known for what separates them than for what brings them together. Perhaps the only thing the eight countries of the Gulf really share other than sand, oil and heat, is the sea.

The lands that surround the Gulf contrast sharply with its waters. Barren land is met by a very productive sea. The region may be energy-rich, but it is resource poor in every other sense, except fisheries.

As the beginning of the world's largest oil highway, the Persian Gulf is perhaps the most strategic body of water in the world. It is an important part of the world's energy life-line. Yet the Gulf is the region's life-line in other ways too. Most of the Gulf's population is squeezed along the coast and it is a source of food and drinking water.

Recognizing that the health of the Gulf is a common concern throughout the region, the eight Gulf States, in co-operation with UNEP, launched the Kuwait Action Plan in 1978.

Geography of scarcity

An old Arab proverb from the Gulf says "the sea is life, the sea is sustenance. Without it we perish under the sand." With a geography dominated by sand, this statement could not be more true. Summer temperatures regularly rise above 40° Centigrade, making the Gulf one of the hottest places on earth. Asphalt roads turn into mush, concrete crumbles, water evaporates in minutes, buildings need constant repair from heat damage. It is also one of the driest regions on earth. The Arabian Peninsula receives less than 10 cm of rainfall a year.

The Gulf itself is not very big. It stretches only 1,000 km from the Shatt Al Arab waterway in southern Iraq to the Strait of Hormuz, and varies in width from 75 to 350 km. With an average depth of only 35 m, it is considerably shallower than the Mediterranean or Red Seas. On the other hand, strong tidal currents rapidly flush out the Persian Gulf. Its total water volume is completely replaced every 1–3 years. This fortunate hydrobiological feature means that oil and other pollutants from domestic, industrial and agricultural activities are pushed out into the much deeper and larger Arabian Sea.

Around 10 million people live along the Gulf's highly developed coastline. With the exception of Iran and Iraq, nearly all of them are packed into the coastal zone, bolstered by some 5 million migrants and foreign "contract workers". Population growth is also concentrated along the Gulf. With the one exception of Iran, all the Gulf countries' populations are growing by more than three per cent a year, doubling their numbers every 20 years.

This very growth threatens to undermine the region's economic stability. For the moment, Saudi Arabia grants every citizen a plot of land and an $80,000 government loan to build a house on it. There are no income taxes and all public services are free. By the twenty-first century, these luxuries may have to be cut.

"The entire coastline of Kuwait is packed with people," observes Dr Manaf Behbehani, professor of marine science at Kuwait University. "The pressures on our coastline are enormous. Nearly all new construction is along Kuwait Bay: offices, homes, resorts. And this urban expansion has ruined tidal flats and wildlife habitat. The shoreline of Kuwait City has actually been extended 60 metres into the Bay. Our shoreline is expanding at the expense of coastal resources." The same story is heard throughout the entire western and southern rim of the Gulf. Oil wealth has fuelled a building boom – everything from flats to industrial complexes – which has outpaced government efforts to control it. Although Kuwait City has a master plan for phased development, many other cities in the region do not. Some cities have doubled their populations in as little as four years.

Kuwait now has a development plan for its entire coastline. But land reserved for oil and gas production, and other industrial uses, accounts for over 50 per cent of it.

Fresh water is a scarce commodity in the Gulf and more valuable than oil. Kuwait, like the other Gulf States, must get most of its drinking water from the sea. Five big desalination plants produce 148 million litres of potable water per day. This is produced by mixing distilled water produced by the plants with brackish water from coastal aquifers. A sixth plant went on stream in 1990.

In all the Gulf there is only one hydraulic laboratory which studies the effects of coastal engineering on tides, currents, and marine organisms: the Hydraulic and Coastal Engineering Department of the Kuwait Institute for Scientific Research (KISR). Kuwait, like its neighbours, is hobbled by a lack of co-ordination when it comes to coastal management. There is not one government department charged with overall responsibility for managing the coastal zone. Instead, management is fragmented between several competing ministries and research departments. Too often the result is coastal chaos, rather than coastal management.

Coastal pollution

The Gulf States have transformed themselves from medieval to industrial economies in a mere 40 years. But such rapid, often unplanned, urban and industrial growth has also resulted in severe coastal pollution and loss of wildlife habitat. Millions of tonnes of untreated industrial wastes are discharged into the Gulf every year. Oil pollution is even more serious.

All of the Gulf States have working sewage treatment plants, covering most of their coastal populations. Nearly 95 per cent of Kuwait's population is now served by three-stage sewage treatment plants, which process 80,000 cu m of sewage water per day. Similarly, 74 per cent of Bahrain's population is covered by sewage treatment networks, as is 96 per cent of the United Arab Emirates', and 79 per cent of Iran's.

By contrast, most Gulf States have done little to control industrial pollution. Kuwait is in the process of building a mammoth industrial waste water treatment plant for its big industrial zone in the south near the city of Al Fuhayhil. The complex is expected to be operating within three years.

In the meantime, coastline investment in the Gulf is estimated to be between $20 million and $40 million per km, with 20 major industrial centres completed or under construction. So far, few states have well-established integrated pollution control programmes.

On the other hand, controlling industrial and municipal pollution is receiving increased attention in the Gulf, if for no other reason than most of the region's drinking water comes from the sea. The intake valves for all desalination plants are monitored carefully for dangerous pollutants. A chlorine plant on Kuwait Bay was closed down when scientists discovered it was discharging mercury into coastal waters. Another problem is jellyfish, which clog the intake pipes of desalination plants and thermal power stations (which use seawater for cooling purposes). Although there is no clear proof that jellyfish swarms are linked to pollution, circumstantial evidence points in that direction.

Red-tides in the Gulf have also increased in recent years, badly affecting fisheries, particularly maricultured species like groupers and sea bream. Although these killer algae have been around for centuries, their outbreaks are aggravated by untreated industrial and municipal wastes which consume oxygen needed by marine organisms. Throughout the Gulf, coastal waters near industries and cities are increasingly oxygen-starved.

Oil: a curse of riches

The Persian Gulf is the most crowded oil highway in the world. Before the war between Iran and Iraq spilled over into the Gulf, some 100 tankers a day sailed through the Strait of Hormuz. Currently, 25,000 tankers sail in and out of the Gulf every year. Around sixty per cent of all oil carried by ships throughout the world – a billion tonnes a year – is exported from 25 major oil terminals scattered around the region. Most oil is carried by tankers at least as large as the quarter-million tonne *Amoco Cadiz*. Before the "tanker war" (1984–8), there were over 800 offshore oil and gas platforms operating in the Gulf as well.

With all this oil being pumped and transported, the Gulf's waters are heavily contaminated with oily residues and tar balls. Around one million tonnes of oil is dumped into the Gulf's waters every year from the routine discharge of dirty ballast water and tanker slops. "Oil pollution is our biggest problem," states Dr Hosny Khordagui, a research associate at KISR in Kuwait City. "Oil pollution affects a lot of our marine biota, tainting fish and mussels." Near shipping lanes and offshore oil rigs, petroleum residues 100 times higher than in uncontaminated waters have been found in the tissues of mussels. Fish are also affected. One KISR scientist states that "fish so far show no signs of pollution stress from hydrocarbons, but every fish we have sampled contains oil in its tissues. Fortunately the levels are too low to raise health concerns."

The eight-year war between Iran and Iraq did not help the situation. When the conflict began to embroil the Gulf itself, in what became known as the "tanker war", during a four-year period 545 tankers were hit by missiles, bombs and mines. "But we were fortunate again," explains Dr Badria Al-Awadi, co-ordinator of the Regional Organization for the Protection of the Marine Environment (ROPME), based in Kuwait. "Despite all the shooting and bombing many of the stricken tankers were on their way into the ROPME sea area when hit, so they were empty. Others managed to contain the oil spills before they spread." Nevertheless, there are untold kilometres of beaches under black oil, oxidized and baked to pavement consistency by the sun.

Besides luck, the flushing capacity of the Gulf and its warm waters help protect it from oil pollution. "As a consequence of repeated oil spills, most beaches in the region are heavily contaminated with tar," observes Olof Linden, a senior scientist at the Swedish Environmental Research Institute in Stockholm. "However, the background levels of petroleum hydrocarbons in sediments and biota are not exceptionally high. This is probably due to rapid degradation and weathering in the Gulf's warm, shallow waters." And to those wonderful tidal currents, which sweep pollution out of the Gulf.

Fisheries: abundance from the sea

The Persian Gulf is endowed with rich marine life. Unlike the Gulf's human population, fisheries benefited from the fighting. Reduced fishing activity in the Gulf gave fish and shellfish stocks a chance to recover from years of over-exploitation. The autumn of 1988 was one of the best years on record for shrimp landings by Kuwaiti and Saudi fishermen. One species of shrimp (*Penaeus semisulcatus*), accounts for 90 per cent of all trawl catches. In the month of September 1988, 2,800 tonnes of shrimp were taken in Kuwaiti waters alone.

Fish stocks have also increased. Groupers, sea bream and snappers are caught with fish traps and gillnets throughout the Gulf. Fish consumption averages about 8 kilogrammes per person per year in the Arab Gulf States. With the war over (for the moment) seafood consumption is on the increase.

Most Gulf States have established fishermen's associations which regulate the number of fishermen and boats allowed to exploit fish and shellfish stocks. In most countries, fishermen need a licence before they can operate, usually controlled by the fishermen's union. Altogether there are probably around fifty commercial operations and about fifty-five artisanal fisheries in the northern part of the Gulf, involving roughly 1,100 boats.

Both Kuwait and Saudi Arabia have imposed bans on shrimping from 1 March to 1 September, giving the shrimp time to spawn. No commercial trawling is permitted in Kuwait Bay, within three miles of shore. This allows artisanal fishermen who specialize in trapping fish in tidal areas a chance to make a better living.

Some Gulf States, like Kuwait, are also developing large-scale fish-farming enterprises for the domestic and export markets. At KISR's Mariculture and Fisheries Department, three species of fish – sea bream, grouper and tilapia – are being experimentally farmed in Kuwait Bay. Sea bream in particular show promise for commercial production in large sea cages. At the moment, KISR researchers have achieved a 12 per cent survival rate for sea bream fry, but hope to be able to increase this to 20 per cent. Groupers are more difficult to farm since they take longer to reach marketable sizes and don't take well to crowded cages. Although stocking rates are lower that with sea bream, the higher prices paid for groupers still make them profitable.

Perhaps the most novel and promising development that KISR researchers have come up with is combining sea bream and tilapia in one big farm, with separate cages. It is more economical to farm two species instead of one. "We have managed to develop tilapia which grow in seawater and brackish water," explains research scientist James Bishop. "Combining sea bream with tilapia has great commercial potential."

Marine conservation

The Gulf War, although beneficial for fish, has not helped conservation efforts. Scientists are still unable to co-operate on Gulf-wide research projects aimed at setting up marine parks and protected areas. Extensive coral reefs off the coast of Saudi Arabia should be protected from exploitation and pollution, but so far nothing has been done to set up reserves. Kuwait is considering a proposal to establish a few marine parks around several of its coral islands, but no action has been taken to implement the plans.

Meanwhile, both habitats and marine life suffer. In Kuwaiti waters there is only one small undisturbed island left where marine turtles can nest; efforts are just getting underway to declare it a sanctuary. Coral reefs, seagrass beds and mangroves are all under stress from coastal development, land reclamation and channel dredging. Land reclamation and dredging, in particular, churn up bottom sediments which kill off seagrass meadows and coral reefs. Island reefs are damaged by boat anchors and trawling operations. About forty per cent of Qatar's coral reefs and nearly a quarter of those around the coast of Bahrain have died, killed by sedimentation and pollution.

Between August and October 1986, thousands of dead and dying dugongs, dolphins, turtles and fish washed up on beaches along the coasts of Qatar and Saudi Arabia. Scientists still don't know what killed them. But industrial pollution and oil are at the top of the suspects' list.

There seems to be little concept of conservation in the Gulf States. In such a harsh environment there may be little to conserve, but what wildlife survives in the desert is often hunted for sport. Bird watchers in other countries complain that migrating birds, particularly raptors and storks, are slaughtered by the thousands on their way through the Gulf region on their way to wintering grounds in Africa.

"At Al Khiran in southern Kuwait, a sea inlet harbors a unique colony of pre-Cambrian blue-green algae known as *Stromatolites*," states Dr Manaf Behbehani. "These animals are 2 billion years old – among the oldest living things on earth – yet the government has done nothing to protect the area from development."

Every spring the Environment Protection Council of Kuwait launches a public information campaign in an attempt to alert people to the need for conservation of desert habitats and the animal life they shelter. "Now that the war is over," states one Council official, "the Gulf environment is being re-discovered."

The Kuwait Action Plan

The eight governments of the Gulf adopted a Convention and Action Plan in 1978. Two legal instruments – the Regional

Convention, and a protocol concerning "Regional Cooperation in Combating Pollution by Oil and other Harmful Substances in Cases of Emergency" – were ratified and came into force in 1980.

The United Nations Environment Programme acted as interim co-ordinator for the Action Plan until the Regional Organization for the Protection of the Marine Environment (ROPME) began operating on 1 January 1982. The ROPME secretariat was set up in Kuwait.

In connection with the protocol concerning regional co-operation in combating oil spills, the Marine Emergency Mutual Aid Centre (MEMAC) was established in Bahrain in August 1982. "MEMAC's objective is to help contracting states develop their own national capabilities to combat pollution from oil and other harmful substances as well as to co-ordinate information exchange, technological co-operation, and training," points out Hamid Shuaib, President of the Kuwait Environment Protection Society. "MEMAC took a leading role in combating oil pollution during the Naw Roz oil-field disaster in 1983 and the Assimi oil tanker disaster in 1985."

A second protocol on controlling oil pollution, this one designed to combat pollution from offshore activities, was signed in December 1988. And two more protocols are in the works: one dealing with the control of land-based sources of pollution, and another concerning the transport of hazardous substances.

During the initial phase of the Action Plan, the Gulf States concentrated on oceanographic studies and baseline pollution surveys. A major emphasis has been on making data comparable with information gathered in other regional seas programmes. Many technical specialists were also trained. Since 1982 more than 500 technicians have received special training in oil and non-oil pollutant sampling, data handling, oceanographic modelling, marine monitoring and research, and marine pollution prevention.

The Action Plan as a whole focuses on oil pollution, industrial wastes, sewage treatment, fisheries resources, and the environmental impact of coastal engineering and mining. "The Gulf is like a big lake for us," states Dr Badria Al-Awadi, of ROPME. "Now that the war is over, our biggest priority for the next five years is the management of coastal resources, particularly fisheries."

The Iran–Iraq War has dominated the region for eight years. Now that peace has come, perhaps the Gulf States can start co-operating on a regional level and begin to use the Action Plan for the betterment of the Gulf environment. Even during the darkest hours of fighting, scientists from both Iran and Iraq still attended ROPME meetings. In late October 1988 an emergency meeting of ROPME was called in Kuwait, attended by representatives from both Iran and Iraq. At the top of the agenda was a plan to deal collectively with hazards to Gulf shipping posed by the leftovers of war: unexploded mines and other munitions, and the wreckage of ships sunk during the tanker war. Perhaps if the Gulf States can co-operate on this level, they can also co-operate on regional security matters.

Unfortunately, for the most part the war has made the environment an issue of secondary importance. And ROPME has suffered from a general lack of interest in its programmes. The oil-rich states of the Gulf have handicapped ROPME with a budget of only $1.5 million a year, barely enough to pay for basic co-ordination work. Yet ROPME remains the only truly region-wide forum in the area through which co-operation on marine and environmental issues can be achieved.

5. The Wider Caribbean

The Caribbean is an area so diverse that it defies description. "Races of all continents have miscegenated here, cultures and customs most foreign to each other have syncretized, nature has exploded into myriads of forms, and the constant feast of color and sound has goaded the formal carnival to paroxysm" writes literary critic Ramon Mendoza in the *Caribbean Review*.

Yet behind this carnival of abundance there is scarcity – of jobs, skills, materials, and money. And in many parts of the Caribbean, resource abundance is giving way to degradation and over-use. Despite frequent development initiatives like former President Reagan's proposed "mini-Marshall Plan" for the region, poverty is now more widespread than ever.

Diversity – economic, ecological and cultural – brings its own set of problems. This bizarre assortment of tropical and sub-tropical ecosystems, encompassing 24 island states and territories and 12 mainland nations, has long been divided between the Hispanic group of countries and the separate English-, French- and Dutch-speaking states. As a result, communication, not to mention co-operation, has often been rare.

Co-operation has been hampered also by the economic divide which cuts through the region. Sitting on top of the Caribbean is the United States, while the southern rim is dominated by the the rapidly developing economies of Colombia and Venezuela. In between lies the penury of most of Central America and the islands (with a few notable exceptions, such as Costa Rica).

Despite spasmodic attempts at modernization, many Caribbean countries start the development process with severe handicaps. Grappling with the environmental effects of crushing poverty means that hasty development projects often contribute to the problems rather than the solutions. Caribbean ecosystems are beset by problems: chemical pollution produced by industry and agriculture; silt from dredge-and-fill operations and poor land management;

irrational exploitation of coastal resources; untreated wastes from coastal cities and tourist centres.

Realizing that the Caribbean needed a mechanism through which regional co-operation could be achieved in combating environmental degradation and resource impoverishment, 23 of the region's countries and territories adopted an Action Plan for the Caribbean Environment Programme in April 1981. The Caribbean Environment Programme, drawn up by UNEP, remains a landmark in regional co-operation; one of the few forums available for dealing collectively with common environmental concerns.

The Wider Caribbean is not nearly as landlocked as the Mediterranean Sea or the Persian Gulf. It consists of two huge basins: the Caribbean Sea proper and the Gulf of Mexico. Its surface area covers over 4 million sq km. The "American Mediterranean" is also deep, averaging 2,200 m, with the deepest part – known as the Cayman Trench – plunging down to 7,100 m.

The drainage basin of the sea is equally enormous, encompassing some 7.5 million sq km and eight major river systems, from the Mississippi in the Gulf to the Orinoco in Venezuela. These rivers also freight in millions of tonnes of pollutants from the hinterlands.

The Caribbean is the origin of the Gulf Stream. The Caribbean Current and Antilles Current join up at the tip of Florida to form this huge wedge of warm water which then sweeps across the North Atlantic to northern Europe, and finally to Scandinavia and into Arctic Russia, making life in Northern Europe more bearable.

Most of the 22 islands states and territories are very small, with more people than the land can support. Collectively, they comprise a mere five per cent of the region's total land area and hold only 15 per cent of the Caribbean's population. Yet, with the exception of Cuba and the Dominican Republic, population densities on the Antilles islands exceed 100 people per sq km; Barbados has 550 people per sq km.

Small may be beautiful, but too many Caribbean states, especially the islands, must somehow subsist on one or two primary export commodities. Jamaica's main export is bauxite, Cuba's sugar, Costa Rica's bananas, Trinidad and Tobago's oil.

Another consequence of too little land and too many people is the necessity to import food. Most of the Caribbean islands even have to import seafood. In Dominica and St Kitts local demand for fish exceeds supply by more than 250 per cent, while imports of fish to St Lucia were valued at more than $2 million in 1982.

Squeezing the coasts

Most of the smaller islands in the Greater and Lesser Antilles consist of nothing but coasts. But even along mainland Central America and the Caribbean coast of Colombia and Venezuela, human numbers are rapidly outstripping resources. Most of the Basin's 170 million permanent residents (including those along the Gulf coast of Mexico and the United States) live on or near the seashore. They are joined every year by 100 million tourists, nearly all of whom end up at the coasts. The Caribbean has become one of the world's most popular holiday destinations (second only to the Mediterranean).

Alfred Taylor, a native Barbadian and president of the Caribbean Hotel Association, says:

> Tourism is the future of the Caribbean. But at the same time, we have to be very careful about our environment. Solid waste disposal is now a serious problem. Our waters are getting more polluted. Our reefs are dying. On a lot of islands the hotels are too close to the beaches. The sewage pollution is killing the reefs, which then causes beach erosion. If we are not careful, we will end up with loads of hotels, but no beaches and tourists.

The environmental effects of coastal crowding are obvious. Mangroves give way to squatter settlements and shanty towns or are cut down for timber or to make room for tourist resorts. Coral reefs are over-fished and over-exploited for building materials and coffee table trophies. Without proper regulations and zoning restrictions, coastal development spreads unchecked. And population growth rates continue to hover between two and three per cent a year for many of the poorer Caribbean islands and mainland states. This means that for countries like Haiti, Honduras, and the Dominican Republic, already struggling to balance human needs with dwindling resources, their populations double every 23–35 years.

Demographic trends remain unsettling. During the past three decades the growth of urban centres has continued unabated. Most of it has occurred in medium-sized and large cities; those with populations of 100,000 or more. This uneven distribution of population works against attempts at managing resources sustainably. Since most of the new migrants to urban areas are poor peasants forced off exhausted land no longer capable of supporting them, they join the ranks of the unemployed and under-employed, in squatter settlements and shanty towns. Here they contribute to the problems municipal governments face in attempting to provide basic services such as potable water, sanitation facilities, health care and education. In most cases beleaguered governments cannot hope to match services with ever-increasing numbers of urban poor. Hence, the conditions of most urban slums and squatter settlements continue to deteriorate. And so too does the state of coastal resources, overwhelmed by numbers and needs.

Diarrhoeal diseases caused by water pollution and contaminated food are among the leading causes of death in the region, particularly of children under five years of age. In the late 1970s, diarrhoeal diseases in Guatemala and Nicaragua accounted for 26 per cent and 34 per cent, respectively, of all the deaths registered.

Life expectancy varies considerably too. The average Haitian can expect to live only 51 years, compared to 74 years for Puerto Ricans and 75 years for Gulf Coast Americans.

Coastal pollution

As coastal populations increase, resources suffer and pollution mounts. In 1979, UNEP studies showed that the wastes from at least 30 million people flowed into the Caribbean's coastal waters without treatment of any kind. Today, less than 10 per cent of the wastes generated by the Basin's 170 million residents receive any form of treatment before being dumped in coastal waters or into rivers which end up discharging their loads into coastal areas. On top of this comes the untreated wastes of the tourists.

The waters along nearly every urbanized coastline are clogged with raw sewage and municipal garbage. In some cases – as on Haiti – health alerts have been issued, and bathing beaches closed. There have even been sporadic outbreaks of cholera and typhoid, and pollution-induced diarrhoeal diseases are endemic throughout much of the region. The near-shore waters around cities like Port-au-Prince, Havana, Kingston and San Juan are so choked by untreated sewage and other municipal wastes that they are becoming oxygen-starved.

Not only municipal wastes contribute to the pollution problems of the Caribbean. The high organic loads from sugar-cane mills and food-processing plants rob shallow coastal waters of oxygen, causing anoxic conditions in which few marine organisms can survive. In the Mississippi Delta south of New Orleans, Kingston harbour, Jamaica, and off the coast of Trinidad, toxic industrial wastes kill fish and destroy marine habitats. In March 1988 a raft of dead fish 1.5 km long and 300 m wide was found in the Gulf of Paria – between Venezuela and the islands of Trinidad and Tobago – victims of industrial wastes and hydrocarbon pollution.

Heavy metals from mining operations and metal smelting pose serious threats to coastal marine habitats in many areas of the Caribbean. In the Coatzacoalcos Estuary in the Gulf of Mexico, bottom sediments and marine fish and shellfish contain high levels of lead, cadmium, mercury and copper. Dangerously high levels of mercury pollution have also been documented for the Bays of Cartagena in Colombia, Guayanilla in Puerto Rico and for Puerto Moron in Venezuela. Havana Bay is full of heavy metals, including mercury, lead, and cadmium.

The widespread use of agro-chemicals and pesticides contributes another set of pollutants to the marine environment. Like heavy metals, these persistent poisons lodge in sediments and bio-accumulate in fish and shellfish. Around Puerto Rico and the islands of the eastern Caribbean, measurable levels of pesticides have been recorded in the water column. Pesticide run-off into coastal waters has killed fish around Jamaica and off the coast of Colombia. Years after both were banned from general use, long-lived pesticides such as DDT and DDE have shown up in the tissues of reef-dwelling fish like groupers taken in the Gulf of Mexico and the Grand Bahamas. Shrimp and plankton from the northern Caribbean were found to contain measurable levels of DDT as well, but not in high amounts.

The destruction of near-shore marine habitats – mangrove forests, seagrass beds and coral reefs – by pollution and unplanned coastal development is drawing considerable attention from Caribbean governments, not to mention conservationists, fisheries managers, resource planners and the concerned public. The combined effects of increasing populations and degraded resources are prodding many Caribbean governments into developing coastal area management plans.

Oil pollution

Every day some 5 million barrels of oil are transported through the Caribbean. Predictably, like all seas in which oil or gas is extracted and through which petroleum products are transported, the Caribbean suffers from oil pollution. Every year on average, about 7 million barrels of oil are dumped into the Caribbean. Some 50 per cent of this pollution is thought to be accounted for by tankers and other ships, discharging oily wastes, dirty bilge waters and tanker slops in direct violation of IMO treaties. A significant amount of oil also finds its way into Caribbean waters from offshore oil rigs and exploratory drilling (in 1978, nearly 77 million barrels of oil were released from oil rigs).

The Gulf of Mexico has the distinction of hosting the world's worst oil spill. In the early morning hours of 4 June 1979, the Ixtoc 1 exploratory oil well in the Bay of Campeche blew out. It was finally capped on 23 March 1980, 290 days later. During this time 475,000 metric tonnes of oil were spilled into the warm waters of the Gulf. Although hundreds of thousands of oil-soaked crustaceans were washed up on Gulf beaches for months, the full extent of the damage was never calculated.

The oil-producing countries – Colombia, Venezuela, Mexico, Trindad and Tobago and Barbados – extract oil at the rate of 3.5 million barrels a day. The big three – Colombia, Venezuela and Mexico – have petroleum reserves totalling 12 billion metric tonnes. Not only does the Caribbean produce oil and gas, exporting most of it, but much of it is refined in the region. Across the Caribbean, including the Gulf of Mexico, some 73 refineries are capable of handling over 12 million barrels of oil per day.

To a large extent, oil has fuelled the development plans of Mexico, Colombia, Venezuela and Trindad and Tobago. Another result of the oil boom has been devastated coastal ecosystems in major oil producing areas – the Gulf of Mexico, Venezuela, Trinidad – and along tanker routes. Mangrove swamps, seagrass meadows and coral reefs have been wiped out by oil spills in many parts of the Caribbean. The windward exposed beaches from Barbados to Florida are heavily contaminated with tar balls and oily residues; in some cases as much as 100 grammes of tar have been found per metre of beach front. Oil not only tars beaches and reduces tourist dollars, it kills marine life. Sea turtles are especially susceptible, since they sometimes ingest floating tar and die. And once oil hits coastlines, it can devastate marine communities, killing off entire populations of shellfish and crustaceans and fouling habitats.

Endangered ecosystems

Mangrove swamps, seagrass beds and coral reefs, in addition to being the most biologically productive marine habitats, are also prime nursery grounds for commercially important fish and shellfish. They are all coming under increasing stress throughout the Caribbean from pollution, sedimentation, and the direct effects of dredging and coastal land reclamation. If left unchecked, the destruction of these vitally important habitats could sterilize coastal areas, greatly reducing their productive capacities. Such a development would have drastic effects on poor coastal populations.

Everywhere in the Caribbean, with a few exceptions, mangrove swamps are being degraded and destroyed at unprecedented rates, especially on the islands of the Greater Antillies. Puerto Rico's mangrove forests, for example, have been reduced by 75 per cent since the first Europeans began colonizing the island. Nearly all of Haiti's mangroves have been felled by poor peasants and sold for timber, fuelwood, or charcoal.

Large areas of mangrove forests have been destroyed in the Orinoco Delta of Venezeula, mostly through the opening up of ship channels. Similarly, parts of the Caroni Swamp on Trinidad have been ruined by the construction of channels and port facilities.

Increasingly, deforestation of uplands and coastal areas brings in its train erosion of agricultural lands, coupled to landslides during the rainy season and droughts during the dry season. Rivers draining such area run swollen with sediments. As more soil is washed into coastal areas, seagrasses and coral reefs are smothered and killed.

Seagrasses are particularly susceptible to disturbances. They are the only flowering plants that have returned to the sea, and look very much like their ancestors on land. Their entire life cycle, including pollination, occurs underwater. Some 50 species of seagrasses are distributed throughout both temperate and tropical seas, thriving in clear, shallow waters. Seagrass meadows are an important link in the complicated food chain which ties coastal wetlands such as estuaries and mangroves to offshore coral reefs.

Like mangrove forests, seagrass meadows help collect and anchor sediments, making surrounding waters less turbid. This cleansing ability of seagrasses improves water quality not only for themselves but also for associated communities of filter-feeders (clams and oysters) and nearby coral reefs. Submarine meadows provide a host of organisms with shelter, nurseries, and food. Many fish and invertebrates, as well as sea turtles and dugongs, graze the meadows, eating algae and other plant matter that grow on the surface of seagrasses or in their litter. In addition, because of their strategic position between mangrove wetlands and coral reefs, tropical seagrass communities act as effective buffers, modifying wave action and taking nutrients to and from these other ecosystems.

Seagrass beds share a similar fate, and are affected by some of the same kinds of pressures which destroy mangroves. The greatest threats facing seagrass meadows in the Caribbean (and elsewhere) include dredge-and-fill operations, erosion sediment from coastal deforestation and poor agricultural practices, fishing with bottom trawls, and water pollution caused by industrial and municipal wastes, thermal discharges from power plants, and oil spills.

Unlike mangroves, the magnitude of the destruction of seagrass beds is largely unknown. Seagrass loss has been documented in only a few countries, notably Australia and the United States. One study carried out in Boca Ciega Bay, Florida, following a dredge-and-fill operation to enlarge a boat harbour, revealed that a full 20 per cent of the seagrass community in the bay had been wiped out, causing an 80 per cent reduction in fish species and a fisheries loss estimated at $1.4 million.

Often the trouble starts when mangrove forests are clear-cut. But probably nothing is more harmful to seagrass beds than dredge-and-fill operations. Whether for the construction of harbours, residential estates, coastal industries, or ship channels, dredging churns up enormous quantities of bottom sediment. Water quality is impaired and visibility reduced as suspended particles of sand and mud clog the water column. The resulting turbidity interferes with photosynthesis and reproduction, and when the sediment finally settles, it often buries seagrasses. To make matters worse, dredge-spoils (the "fill" part of the operation) are often dumped indiscriminately over seagrass beds.

In efforts to correct past mistakes and to better manage coastal environments, a number of countries (mainly France, Australia, England, the United States and the Philippines) have launched programmes to rehabilitate seagrass meadows. So far, restoration efforts have been successful with four varieties, including one of the most common species (*Thalassia testudinum*), known as turtlegrass. But the costs of re-introduction are high: ranging from $3,000 to $25,000 per hectare.

Dr Anitra Thorhaug, professor of biological sciences at Florida International University in Miami, has successfully replanted seagrasses in Biscayne Bay, Florida, using student labour. But it is time-consuming and hard work, since the seedlings have to be planted by hand. Thorhaug states that

> seagrasses and the important fish nurseries associated with them have been badly neglected and damaged throughout many areas of the world, particularly in the Caribbean. But efforts are underway to conserve remaining seagrass meadows and to restore others which have been lost. Unfortunately, the combination of seagrass vulnerability to pollutants and their tendency to grow close to shore, where dumping occurs most frequently, has left large parts of the Caribbean denuded of seagrass.

Caribbean forests are rapidly disappearing as well. Every year on average 1.8 million hectares of Caribbean forests are destroyed, while only 34,000 hectares are replanted. According to Norman Myers, author and consultant to various UN agencies, the forested area of Central America and Panama decreased by 112,800 sq km between 1961 and 1978. Much of this forested land was turned into cattle ranches and cash crops. But slash and burn subsistence cultivators and loggers are mostly responsible for the destruction of upland watersheds. In the rural areas of some energy-poor islands like Haiti, Martinique and Guadeloupe, fuelwood gatherers also contribute to the process of deforestation.

In *Bordering on Trouble*, environmental journalist Larry Mosher describes what is happening to the forests of St Lucia, a beautiful volcanic island in the Lesser Antilles.

> Much of its forest loss stems from the monoculture export of bananas. In addition to degrading the island's fragile topsoil, such plantation practices have forced many farmers off the best land and into the hills, where they slash and burn to raise food as well as grow more bananas. This exacerbates soil loss through erosion, which in turn forces the farmers to move on to other areas after several growing seasons when their crop yields diminish. The vicious cycle can ultimately change the microclimate, transforming farmland into semi-desert.

Once the forest cover has been stripped away, hilly areas quickly shed their topsoil. Panama has nearly one million hectares of eroded soils, Venezuela has ten times that amount. Measured rates of soil loss in the Caribbean are as high as 35 tonnes per hectare. Soil stabilization and reclamation measures could rehabilitate some of this degraded land, but the costs are usually prohibitive.

The alternative to rehabilitation, however is often far worse. "Once you've lost the hills, you lose the sea," points out Beverly Miller, Senior Programme Officer in the Regional Co-ordinating Unit of the Caribbean Environment Programme, based in Kingston, Jamaica. "We are making the critical links between what happens on land and how this affects what goes on in coastal areas." Coastal-zone management plans now call for reforestation of the uplands in many areas to guard against further degradation of near-shore resources.

Fisheries

The Caribbean does not have extended continental shelves stretching out from land masses and the islands. Because of a pronounced lack of upwellings of nutrient-rich subsurface water, surface water is generally poor in nutrients. Consequently, mangrove forests, seagrass meadows and coral reefs play major roles in providing critical breeding and nursery habitats for many species of fish, lobsters, crabs, mussels and oysters. Seagrass beds alone account for around eighty per cent of the breeding grounds for a wide variety of fish.

Caribbean fisheries are confined largely to smaller-scale commercial operations and artisanal activities. Nevertheless, the Inter-American Development Bank reported that Caribbean fisheries landed some 9 million tonnes in 1980, worth around $3 billion. Altogether, some 2 million people are directly engaged in fishing around the Wider Caribbean.

The richest fishing grounds in the Caribbean are found on the Campeche Bank in the Gulf of Mexico (site of the Ixtoc I oil spill), the Mosquito Bank off the coasts of Honduras and Nicaragua, in th Gulf of Paria (with industrial and municipal pollution) between Venezuela and Trinidad and Tobago, and in the coastal waters of Guyana and Suriname.

Unfortunately, two of these important fishing areas are chronically polluted. The Campeche Bank is filled with offshore oil drilling platforms and is the site of the notorious Ixtoc 1 oil spill. The Gulf of Paria is contaminated with industrial and municipal effluents.

In the last few years, longlining has increased significantly in the deeper waters of the Caribbean. Commercial fleets from Japan, Taiwan and South Korea dominate longline fishing operations. There are fears that this technique might well deplete certain stocks of Caribbean fish, especially tuna and billfish (like the marlin). One UNEP study indicates that a sustainable yield of fish and shellfish should not exceed 2.6 million tonnes a year. If true, then the Caribbean is already being over-fished by around 6 million tonnes a year. In many areas of the eastern Caribbean, fish stocks are becoming depleted. Management is badly needed to get Caribbean fisheries onto a sustainable footing.

Meanwhile, there is great potential for aquaculture and mariculture in the Caribbean. Pond-raised shrimp, prawns, and spiny lobsters are being marketed in Panama, Colombia, Venezuela, Mexico and the US Gulf States. The Dominican Republic has introduced large-scale aquaculture and mariculture operations. Caribbean spider crabs (*Mithrax spinosissimus*) are being raised in ponds. They can grow up to two feet long and weigh two kilogrammes. On tiny Dominica, the government introduced a programme to culture Marron lobsters, a freshwater lobster from Australia that can grow as large as Atlantic lobsters from Maine.

Marine conservation

With so many critical habitats in the Caribbean being destroyed by unplanned coastal development, sediment from deforestation, coastal erosion and dredging, municipal and industrial pollution, and oil spills, it is hardly surprising that most marine mammals and many amphibians and reptiles are now on lists of endangered species.

The Caribbean monk seal (*Monachus tropicalis*) is thought to be almost or entirely extinct. The number of manatees is constantly declining due to habitat loss, entanglement in fishing gear, and collisions with motor boats. Crocodiles and alligators continue to be poached throughout the Caribbean, despite laws against it. Dolphin species are dwindling, victims of fishing nets.

Tragically, virtually all of the region's marine turtles are endangered: loggerhead, green sea, hawksbill, Kemp's Ridley, Central American river, and leatherback. They are running out of nesting areas as their beaches are taken over for tourist resorts, housing, or industries. Adult sea turtles are killed for their meat and shells. In 1974 some 40,000 sea turtles came ashore to lay eggs in the Gulf of Mexico. By 1976 only 700 were found, and a year later the total was a mere 450.

The loss of wildlife habitat and species in the Caribbean is a scandal. It has been estimated that roughly forty per cent of all global vertebrate extinctions have occured in the Caribbean Basin. Many of these extinctions are attributed to the indiscriminate destruction of upland forests and coastal mangroves, and the extensive use (and misuse) of agricultural chemicals.

Coastal erosion

Shore and beach erosion, often due to the destruction of mangrove forests and coral reefs, is a growing menace in Puerto Rico, Jamaica, Trinidad and the States bordering the Gulf of Mexico: Florida, Mississippi, and Louisiana.

Beach stabilization projects are being tried out in Florida and Mississippi, with a certain degree of success. Mangroves are being replanted along the Louisiana coastline, in an attempt to hold back the sea. But clearly the best way to avoid coastal erosion is to prevent it from ever happening – and to do this an effective coastal management programme is needed.

The Caribbean Environment Programme (CEP)

In 1976 UNEP, in co-operation with the Economic Commission for Latin America, established a joint project aimed at developing an action plan for the sustainable management of the Caribbean environment. After lengthy consultations with Caribbean governments, the CEP's Action Plan was finally adopted at a high level meeting in Montego Bay, Jamaica in April 1981.

Priority projects were selected for the first phase of the Action Plan and a trust fund was established to finance the programme. Unfortunately, "for almost three years, the Caribbean Action Plan remained merely a diplomatic vision," writes Larry Mosher. "Its nine member monitoring committee met twice, first in New York City and then in Cartagena, Colombia, mostly to agree that nothing could start until its trust fund grew larger. By mid-1982, a year after CAP's inauguration at Montego Bay, its trust fund had received only $25,355 out of initial pledges that totaled $1.[illegible] million." It was not until the autumn of 1983 that the monitoring committee finally had enough money to begin work.

Their first effort suffered from financial difficulties, but was still impressive. An Intergovernmental Meeting in Cartagena, Colombia, in 1983 adopted two important legal instruments for dealing with common environmental concerns: "The convention for the Protection and Development of the Marine Environment in the Wider Caribbean Region" and a "Protocol Concerning Co-operation in Combating Oil Spills in the Wider Caribbean Region".

The Regional Co-ordinating Unit, set up in Kingston, Jamaica, finally opened in September 1986, five years after the Action Plan had been adopted. Both the Cartagena Convention and the Protocol on oil spills were signed or acceded to by 16 countries and the European Economic Community and entered into force in October 1986.

Over the past 9 years, the Caribbean Environment Programme has received around $8 million, mostly from UNEP's Environment Fund and from contributions by the CEP countries. The United States, by far the dominant economy of the region, has given no money to the Caribbean Trust Fund. Instead, the US donates experts to the Co-ordinating Unit in Jamaica and provides other services in kind. Congressional and State Department politics, revolving around Cuba's participation, have prevented the US from fully participating in the Caribbean Environment Programme, a situation that irritates many of the poor Caribbean states. "If France was giving $375,000, we had hoped for $500,000 from the United States," recalls Arsenio Rodriquez, the UNEP official who helped organize the Caribbean Action Plan. "But because the United States consistently sabotaged the trust fund, the action plan faced a severe financial crisis. It's so petty, just to keep a few dollars from going to Cuba."

Despite of the weakness of the US involvement in the CEP, the governments of the region seem to be taking resource management and environmental issues seriously. "Environmental concerns are now a part of each election in the region," notes Beverly Miller. "In fact, the environment is one of the most important political issues in the Caribbean." In February 1989, Michael Manley replaced Edward Seaga as Prime Minister of Jamaica. He won re-election partly on a green platform, promising better resource management.

One of the reasons environmental and resource issues are finally being considered by politicians is the success of information campaigns carried out by the largest conservation NGO in the region: the Caribbean Conservation Association, based in Bridgetown, Barbados. The awareness fostered by this NGO has led to the demand for action to be taken to solve the environmental and resource problems now facing the Caribbean.

A milestone was reached recently when the CEP countries agreed to initial proposals for the development and implementation of a long-term strategy for sustainable economic growth, through the rational management of the marine and coastal resources of the Wider Caribbean.

Despite chronic budget problems, the Caribbean Environment Programme has managed to implement a number of important baseline studies and has set up institutional mechanisms for co-operation in the region. One of CEP's most successful projects was carried out by the Caribbean Environmental Health Institute; it produced country reports for the island states on all major land-based sources of pollution of their marine and coastal environments. The next stage will involve clean-up programmes and the development of coastal-zone management plans. For now, CEP has made the management of coastal and marine resources, and the assessment and control of marine pollution, urgent priorities.

6. The South Pacific

The island chains of the South Pacific are nothing more than tiny specks of sand, coral and rock strewn across a vast expanse of ocean. Many archipelagos are isolated from each other by hundreds of kilometres of sea. These vast distances are made wider by ethnic and cultural divisions that split the Pacific into three distinct parts: Micronesia, Melanesia and Polynesia. Even within these major groups, island cultures are often at odds with each other over political, economic and social issues. Language barriers can be formidable: in Papua New Guinea alone, some 700 different languages are spoken.

"The picture postcard image of the South Pacific is, in many places, false," asserts Dr John Pernetta, Associate Professor of Vertebrate Biology at the University of Papua New Guinea. "Behind the facade of palm trees waving gently in the breeze is the stark reality of grinding poverty set against a backdrop of some really basic environmental problems."

Among these basic problems are: widespread destruction of mangrove forests, seagrass beds and coral reefs around populated islands and atolls; runaway urbanization; depletion of fisheries resources along densely populated coasts; deforestation of upland watersheds; soil erosion; mis-use of pesticides; pollution of rivers and streams from mining activities; improper disposal of industrial and municipal wastes; lack of sewage treatment facilities; inadequate provision of fresh water; and species loss.

As settlements grew and cash economies developed, many island communities found themselves displaced by the very development that promised a better life. Limited island resources are capable of supporting limited populations. For many Pacific island societies, which had evolved practical and sustainable ways to utilize their resources, Western style development proved their downfall. Too many people pushing upon the coastlines ruin critical habitats for marine turtles and nesting birds, and economically important fish and shellfish. Too many farmers in the hills bring about deforestation and the erosion of soils needed to sustain them, adding to the problems of the coasts. Too much big development, without regard to local needs, more often than not exacerbates the problems instead of contributing to solutions.

As Pacific societies struggle to modernize their economies, resource issues are often neglected. To counter this neglect, and recognizing that small, isolated islands have more in common than just the vast sea around them, the 22 island states and territories of the Pacific have established several mechanisms to deal with collective resource and environmental problems. One of them is the South Pacific Regional Environment Programme (SPREP), initiated with assistance from UNEP, the South Pacific Commission and the South Pacific Forum (through its Secretariat, formerly the South Pacific Bureau of Economic Cooperation).

The largest ocean

The South Pacific region encompasses the largest expanse of ocean in the world, covering about 41 million sq km: nearly twice the size of the Soviet Union. Disregarding Australia and New Zealand, only two per cent of the area is land. Many of the islands are so tiny they have to be exaggerated to be seen on a map.

The distances are daunting. From Guam to Tahiti the distance is 8,000 km, and from the North of the Marianas to Noumea it is 5,500 km. It is 11,000 km from Pitcairn to Palau, equivalent to the distance from Tromsö, Norway to Cape Town, South Africa. From Papeete (Tahiti), the nearest continental landfalls are Sydney at a distance of 6,000 km; San Francisco at 6,600 km; Terre Adelie at 7,000 km; Vladivostok at 7,500 km; and Panama at 8,200 km.

Like the Carribbean, much of the Pacific's waters are nutrient poor. In the surface area of the tropical Pacific, the thin warm upper layer of water becomes quickly deprived of nutrients because of photosynthesis and the removal of organic matter as a result of sedimentation. In this subsidence zone, no upwelling of the richer subsurface water is possible and the most extreme conditions of oligotrophy can be found (oligotrophy is the environment in which nutrient concentrations are low and organic production small). These nutrient poor areas are referred to as "ocean deserts". The very clear water results from the absence of suspended particles and hence is very blue. There are few living organisms. For this reason, fisheries are concentrated around island shelves and coral reefs.

Not only is the South Pacific wide it is also very deep. The average depth of much of the Pacific is around 4,000 m, twice as deep as the Caribbean. The various island chains and archipelagos are sometimes bounded by very deep trenches. The deepest and best known of these is the Marianas Trench in Micronesia. The most impressive part of this trench – the Challenger Deep – is over 11,000 m deep.

Except for the continental islands like New Guinea and New Caledonia, most of the Pacific islands have been thrust up out of the sea by volcanic activity. Some islands still sit atop active volcanoes, whereas others have been worn down over the years and are now covered by coral and sand. By contrast, New Guinea's topography ranges from thick tropical lowland jungle to peaks on the Indonesian half of the island which are over 4,000 m and permanently snow-capped.

The urbanization of the Pacific

The island states of the Pacific have been changed dramatically by their integration into the global economy. As rural economies have collapsed, migrants from the interior have swollen city and town populations. The resulting problems – increasing population density, the spread of shanty-towns and squatter settlements, social dislocation and environmental degradation – are all intensifying. Most of the South Pacific's 5 million people are coastal residents. Even in Papua New Guinea the urban and coastal populations are growing more rapidly than those in the forested uplands. Urbanization is a trend the Pacific islands share with virtually every region of the world. But small islands, with limited space and resources, can least afford the damaging affects of crowded coasts. And this lopsided demographic pattern has resulted not only in coastal congestion, but in the depopulation of many small outer islands and atolls.

Not all of the migrants, however, have moved to urban centres on larger islands. Many have emigrated to the developed rim countries: the United States, Canada, New Zealand and Australia. Auckland, New Zealand has become the largest Polynesian city in the world – in the early 1980s these were around 60,000 Polynesians of Pacific island origin there. Similarly, there are twice as many Niueans in

Auckland as there are on Niue and more Cook Islanders than on any of the Cook Islands. Honolulu has more than 20,000 Pacific islanders (including around 14,000 Samoans), and Vancouver, Canada has more than 12,000 Indo-Fijians.

The reasons for this movement of people, both out of the Pacific and to urban centres in it, is explained by Professor John Connell, Senior Lecturer in Geography at the University of Sydney:

> employment opportunities and services (especially education and health) are concentrated in the urban centres; in the small island states of the Pacific, where manpower and capital are often limited, this urban concentration is inevitable on some scale, hence rural – urban migration inevitably follows. In some cases this movement has not affected either agricultural or marine resource production because it has been a movement of "surplus" population. But the usual patterns of labour migration have entailed substantial production losses so that the development of outer islands and especially atolls has become a major problem in many South Pacific countries. The resulting pressure on coastal land and water resources is often very great.

Despite some successful family planning programmes, as on Fiji, the *average* population growth rate of the South Pacific states continues to hover around 2.5 per cent a year, doubling their numbers every 28 years. A large percentage of the population is under the age of 30. The region is still characterized by high birth rates and low death rates. As a result, populations are growing rapidly throughout the Pacific, especially in Melanesia and parts of Micronesia. Much of the growth in Polynesian populations continues to be siphoned off through emigration to the rim countries.

Fiji, however, has well-organized family-planning programmes which have been able to make a big difference in population growth rates. From a growth rate of over three per cent a year in the 1970s, Fiji's present annual increase in population is down to 1.5 per cent.

The Marshall Islands, on the other hand, are reminders of the problems which unplanned urban growth and overcrowding can create. Because these islands lack resources and employment opportunities, most of the archipelago's population lives into two urban centres: the main city of Majuro, and on the island of Ebeye, residence of the Kwajalein atoll workforce, where US nuclear missile systems and radar are tested. The population density of Majuro is over 6,500 per sq km and on Ebeye it is an astounding 25,000 per sq km. Visitors have described both places as unbearably cramped and polluted with all manner of wastes. Water rationing has been introduced in efforts to conserve what little fresh water the islands have. In the dry months, water is supplied for one hour every second day, making flush toilets useless. Without landfills, most household garbage and human wastes end up in shallow lagoons.

Having declared themselves a republic in early 1982, the Marshall Islands quickly discovered that they had nothing to offer anyone except their land. In May of 1982 the government signed a Compact of Free Association with the United States. In exchange for giving the Americans complete control over military matters and a 15-year lease on the Kwajalein atoll, the US, in turn, agreed to pay some $700 million over a 15-year period. "What this has meant," explains Akio Heine, a reporter for *Pacifica 90 Newsmagazine*, "is that the Marshall Islands have gone from subsistence to subsidy." The subsidies have given the islands an economy, but they have not given the people a more liveable environment.

Throughout the Pacific region the problems are beginning to look similar. Unable to provide basic services – sewage treatment, clean water, medical care – for current populations, overcrowded cities and towns continue to expand. Chronic water shortages are forcing many islands to introduce "water hours". Raw sewage, household slops and untreated municipal and industrial wastes are dumped in coastal waters in ever increasing amounts.

Coastal pollution

The urban revolution in the Pacific has contributed greatly to the degradation of coastal resources and to high pollution levels in near-shore waters. The pace of pollution is accelerating, in many cases faster than efforts to control it.

The worst problem facing 90 per cent of the South Pacific's islands is the disposal of sewage and liquid domestic wastes. There are virtually no sewage treatment plants in working order in the region. Nearly all sewage and municipal waste ends up in shallow coastal waters, where it poses grave risks to human health and the environment. In the last few years several cholera epidemics have been traced to shellfish contaminated with raw sewage.

In the interior of most islands, sanitation facilities are non-existent. Fresh water supplies are often fouled by household wastes and human faeces. Intestinal diseases are endemic. Solid wastes are dumped in coastal mangroves or lagoons, where they create breeding grounds for disease organisms. Few islands have bothered to develop solid waste management plans, and supervised landfills are rare.

The coastlines around nearly every urban centre in the Pacific are polluted with untreated sewage, municipal wastes, and household garbage. "One of the main problems is that too many people in the Pacific have acquired a throw-away mentality," states Professor David Mowbray from the University of Papua New Guinea. "In the old days, the people only discarded biodegradables, now they are throwing away everything from plastics and tin cans to car shells."

Industrial pollution affects only a few of the islands – notably Fiji, New Caledonia, Guam and Papua New Guinea. The water in Suva Harbour, Fiji, for example, is toxic from untreated domestic and industrial wastes. The pollution from food-processing plants, oil storage depots, a cement factory, chemical plants, and households is concentrated because Suva Harbour is surrounded by a barrier reef which restricts the mixing of harbour water with the open ocean. This leaves domestic and industrial wastes fermenting in the lagoon, and also permits the accumulation of sediment brought in from bare hills and badly planned coastal development. Poor subsistence fishermen remain dependent on the highly polluted mud flats and lagoon for their source of food.

Another problem afflicting coastal ecosystems is the disposal of pesticide residues and other hazardous chemicals either directly into coastal waters or into rivers which soon dump their toxic loads on the coasts. Many pesticides which are banned or restricted in developed countries are still in use in the Pacific. Even some of the more hazardous organochlorines, like DDT, are being used. Stored chemicals also present hazards. In Tokelau, a warehouse containing barrels of toxic Lindane was swept into a lagoon during a typhoon; the resulting spill killed off a large area of reef and its marine inhabitants.

The danger from pesticides in the Pacific is out of proportion to the amount used – a mere 1,450 tonnes in 1981, compared to over 800,000 tonnes sprayed on Southeast Asian croplands that same year. Unfortunately, many of the people who use pesticides do not recognize the dangers and often don't understand how to mix or apply pesticides safely. Accounts of pesticide poisoning and contamination of coastal areas are on the increase.

There are frequent disasters with toxic chemicals. In August 1983 a barge carrying 2,700 drums of deadly cyanide, destined for the gold and copper mine at Ok Tedi, capsized near the mouth of the Fly River on the south coast of Papua New Guinea. The drums, each containing 102 litres of the poison, were never recovered and the effects of the spill on the ecology of the river mouth are still not known. In another incident at Ok Tedi, the effects were all too evident. When a worker at the ore treatment plant left a valve open, 270 tonnes of cyanide were released into the Fly River. Thousands of fish and crustaceans and scores of salt water crocodiles and water birds were killed. In the early 1980s, subsistence fishermen in the Cook Islands were using highly toxic Dieldrin to kill fish, which they then sold in the local market. The practice was discontinued after several people became seriously ill.

Two of Papua New Guinea's major rivers, the Fly and the Jaba, are routinely polluted with sediments and heavy metals from mining operations. The Jaba River, in particular, is so full of sediments and heavy metals from the Bougainville Copper Mine that its slate-grey waters are completely dead. Researchers claim that wading into the river to take samples is like inching through moving mud. Once the toxic contents of these rivers flow into coastal waters, little is known of how they might contaminate marine organisms, or enter food chains.

In the gold-mining region of Bulolo in eastern Papua New Guinea, miners use mercury to separate gold from the ore. Some of them are beginning to show classic symptoms of mercury poisoning: disorientation and tunnel vision.

As more people migrate from island interiors and from distant atolls to population centres, coastal pollution worsens. Many of the new arrivals find themselves confined to over-crowded shanty-towns and squatter settlements, usually built on the most degraded land. The Koki Settlement in Port Moresby, capital of Papua New Guinea, was built on stilts over the remains of a mangrove forest. Now that there is no forest to blunt storm damage, the poor people of Koki take the brunt of storms themselves. The water around the settlement reeks with garbage and human excrement.

Endangered resources: upland forests, coral reefs and mangroves

Throughout the Pacific, nearly all of the inhabited islands with hilly and mountainous interiors need to conserve vital watersheds. Over-logging in the Gogol Valley of Papua New Guinea might cause increased erosion and landslides (which can occur even on heavily forested slopes) in nearby watersheds. On Fiji, uplands have been cleared for mono-cropping, but logging remains a problem in some areas. Small-scale farmers who strip away vegetation in the hills are frequently rewarded with the erosion of their garden plots during tropical storms. A quarter of Fiji is very degraded land brought on by deforestation, agricultural development on steep slopes and town and city development along the coasts. "The problem is, no one is taking land degradation as a serious environmental problem," complains John Morrison, Director of the Institute of Natural Resources at the University of the South Pacific in Fiji.

Mining for nickel, gold, copper, cobalt, lead, zinc, and chrome has left huge scars in New Caledonia's landscape and caused considerable deforestation over the entire area of the main island, especially in the southern part, where rivers run thick and red with sediments scoured from open-cut mines. At an open-cast nickel mine outside Noumea, a sign in French reads: "Here the earth bleeds". Agricultural activities and the expansion of urban centres have contributed to the destruction. Many of New Caledonia's endemic pines (there are 43 species), decimated from mining operations, are now threatened by logging and coastal development.

Small, heavily-populated islands like Truk, one of the Federated States of Micronesia, are especially hard hit by development. Upland soils are eroding rapidly and coastal areas are being cleared for urban housing and small-scale industrial activities. With 45,000 people on 127 sq km of land, the island's environment is in jeopardy.

Deforestation of uplands and coasts causes the sedimentation of shallow lagoonal and coastal waters. Sediment chokes mangrove forests and smothers coral reefs. This is now a universal problem throughout the South Pacific, wherever coasts have been developed and forests cut down. On New Caledonia, mining and logging in the hills allowed tonnes of fine-grained laterite clay to be carried to coastal waters, where it killed mangroves and filled in estuaries. There are practically no undisturbed mangrove stands left on the entire island.

Fiji has destroyed over 4,000 hectares of mangrove forests for agricultural expansion, mostly to increase sugar cane production. "The irony of the situation," insists Padma Narsey Lal, a Fijian researcher at the East–West Center in Honolulu, "is that, for the most part, mangrove land destroyed in the name of agricultural development has still not been put to use; and where it has been, yields have been very low. This is due in part to acid sulfate conditions in the soils, and the absence of any one authority responsible for the administration of coastal lands."

There are large areas of coral reefs and atolls in the South Pacific – by some estimates as much as 77,000 sq km. Unfortunately most reefs located near inhabited areas are degraded to one degree or another. Coral reef conservation has become a priority for many of the region's countries and territories.

Some of Fiji's reefs have silted over, victims of upland deforestation, the destruction of coastal mangroves for agricultural development, and expansion of coastal towns and cities. On the Solomon Islands, large-scale deforestation and mining activities have ruined near-shore shellfish beds and coral reefs. The pace of upland destruction threatens to get worse; there are now 50 mining companies prospecting for minerals.

As explained in the Introduction, the biggest immediate threat facing Australia's Great Barrier Reef is the crown-of-thorns starfish (*Acanthaster planci*), the coral polyp's natural enemy. Fiji, Western Samoa and Tahiti have suffered extensive coral damage from this predatory animal as well.

Another threat facing the Great Barrier Reef is land-based development. The destruction of Queensland's semi-tropical and tropical forests continues unabated, despite increasingly vocal protests. Much of the coast is already denuded of forest cover and coastal mangroves are now being cut down to make way for urban expansion and tourist centres.

The Queensland coast abounds in classic examples of "counter-development". The town of Cardwell is the main staging area for all tourists who want to camp on Hinchinbrook Island, recently declared a national park. The only way to cross the three kilometres of sea that separate it from the mainland is by boat. In order to make room for the increasing numbers of visitors who want to visit this undeveloped island, the local government in Cardwell decided to clear a stand of mangroves. The predictable result: tonnes of erosion sediment, once trapped by the mangroves, are now heading seaward towards the coast of Hinchinbrook Island. To make matters worse, logging concessions in the watershed around the town may condemn more coastal ecosystems to death by sedimentation. There will be fewer tourists and less money for the town when the reef is dead. In the end, who, except the loggers, really benefits?

Fisheries

Traditionally, Pacific islanders have derived about ninety per cent of their protein from the sea. Many continue to do so. Most of the fishing is artisanal and small-scale, confined to reefs and lagoons. A wide variety of techniques are used, depending on the region and the species of fish to be caught. Fishing craft range from the simple rafts and dug-out canoes with single outriggers used by Polynesians and Melanesians, to the larger double outrigger canoes employed by Micronesians and Papuans. Fishtraps are used extensively in Melanesia; Fijians use nets as well. Normally, fishermen take as much as they need for their own use and sell the rest to neighbouring villages and towns. Artisanal fishing is highly evolved and based on principles of sustainable yield. By obtaining seafood from different habitats – fish from lagoons, shellfish and seaweed from exposed reefs, crabs, lobsters, and molluscs from mangrove swamps, land crabs from the coast, and freshwater prawns from rivers – Pacific islanders are able to diversify their food sources.

Throughout some areas of the Pacific – particularly Melanesia – fishing villages often specialize in different species of fish and shellfish. Since people from one village who want to fish in the waters of a neighbour must first get permission, over-harvesting is usually controlled. When stocks are low, limits are placed on the amount of fish outsiders can harvest.

In New Caledonia, Kanak fishing villages (established on tiny coral islands and islets between the extreme northern end of the main island and the Belep Islands) have evolved some unusual ways to ensure the sustainable use of their coastal resources. Clans are organized into independent management units called a *Kavebu*. There may be more than one clan in each. Besides being a social organization, each *Kavebu* has a well-defined land and marine territory, within which members can harvest freely to satisfy their own needs. *Kavebu* are barred from utilizing each others' territory unless permission is granted. Often, a *Kavebu* will specialize in catching one or two varieties of fish or shellfish. "Techniques are not monopolized," points out social anthropologist Marie Preston, "if a particular fish is found in their *Kavebu*, then they think it is special for them." Some take mostly clams and sea-cucumbers, while others harvest lobsters and rabbitfish.

"Generally, their system works very well," says Preston, who has studied these remote Kanak fishing villages for four years.

"The islands are dry, so they don't have much agriculture. And their fishing is almost exclusively subsistence. Problems arise when outsiders try to fish in their *Kavebu*." A Tahitian who wanted to net fish within a coral lagoon controlled by a *Kavebu* was sent away. In another case, a professional fishing boat from Noumea, filled with weekend anglers, had all their gear confiscated by angry Kanaks, after they tried to fish in a *Kavebu* without obtaining permission.

Although this largely subsistence culture seems to manage its resources well, many of the men have left to join the cash economy on the main island. The women, who do more and more of the fishing, have decided to introduce a small-scale industry of their own: making buttons from trochus shells. In the early part of this century, village women prospered from gathering trochus shells, but they were quickly over-harvested and the industry collapsed. Now, in an effort to re-introduce this source of income, trochus shells are being cultivated for the button trade.

Around other islands subsistence fisheries are in trouble. The Cook Islands and Palmerston Island face the unexplained decline of parrotfish on their coral reefs. On Kiribati, fish stocks are declining and over-fishing is becoming a problem.

This need not happen. If harvested rationally, reef fisheries can supply a tremendous amount of protein. Reef fish around Palau appear to have a potential sustainable harvest of between 2,000 and 11,000 tonnes a year, comparable to the offshore tuna catch. Fish and shellfish from well-managed South Pacific reefs may provide subsistence and small-scale commercial fishermen with 100,000 tonnes a year.

In some areas of the Pacific, fishing is being transformed from a subsistence activity into big business. As the island states and territories of the South Pacific struggle to modernize their feeble economies, a number of indigenous commercial tuna fishing operations have developed. The Solomon Islands, American Samoa and Fiji all have their own tuna fleets, which compete with the deep water fleets from Japan, Korea, the Soviet Union and the United States.

Exclusive Economic Zones (EEZs)

With the advent of United Nations Convention on the Law of the Sea, nearly all of the island states and territories of the South Pacific will benefit immensely from the declaration of 200-mile

Exclusive Economic Zones (EEZs). Much to the consternation of those rim countries with deep water fishing fleets, this development has reduced the area of open ocean in the South Pacific by 30 per cent. At the same time, it has given the small island states and territories of the Pacific a new voice in resource management, and a new source of potential income. Many of the islands are now selling fishing rights within their EEZs to foreign fishing fleets. Kiribati, for example, has sold fishing rights to the Soviet tuna fleet for $1 million a year. Vanuatu has concluded a similar deal. Others are denying access until resource inventories can be completed.

Still, the tuna fleets of Japan, the United States, and Korea, in particular, continue to fish within declared EEZs, since most of the Pacific island states have no way to enforce their claims. This may change if the Law of the Sea Convention can be sorted out to the satisfaction of both the United States and the European Economic Community.

Hazardous wastes in the South Pacific.

At a meeting of South Pacific countries in Noumea, New Caledonia in June 1988, representatives discovered that they had something in common besides the long agenda of shared environmental problems. Within the previous six months, they had all been approached by a US-based company seeking to peddle hazardous wastes from America. On the verge of signing a deal with the waste firm, the King of Tonga was talked out of it at the last minute by his environment representative. Despite the offer of contracts worth millions of dollars, none of the poor Pacific nations approached wanted to make money by taking on imported toxic wastes and the environmental damage they might inflict. The ensuing discussion strengthened their resolve to resist efforts by the developed countries of the Pacific rim, especially the United States and Japan, to foist their toxic waste onto the emerging nations of the Pacific.

Most South Pacific island states have been approached with offers to store everything from poisonous chemical residues to radioactive wastes. The Japanese want to dump low-level radioactive wastes in the Marianas Trench, but will not be allowed to until after the 1990 moratorium on ocean dumping expires. What happens then is unclear. The Japanese have around 600,000 containers of radioactive wastes from their nuclear reactors. So far, they have not been able to ship them somewhere else, and they may end up disposing of their toxic wastes on their own land.

In the early 1980s the US announced plans to store 10,000 tonnes of highly radioactive wastes from Japan, Taiwan and South Korea on three of their Pacific islands: Palmyra, Wake and Midway. When the environmental impact study was completed for this scheme, however, the plans were quietly shelved. Another plan by the US Navy to scuttle 100 out-dated nuclear submarines in the Pacific was abandoned after US environmental groups protested.

Realizing that they are easy targets for other people's hazardous wastes, the countries of the South Pacific took a strong stand on the issue at Rarotonga in 1982, where they gathered to launch a regional environmental programme. Their concerns were translated into the Rarotonga Declaration, in which the conference declared that "the storage and release of nuclear waste in the Pacific regional environment shall be prevented," and "testing of nuclear devices against the wishes of the majority of the people in the Region will not be permitted." Although these declarations are largely rhetorical and unenforceable, their intentions are admirable.

Another issue related to imported hazardous wastes is that of the continued French testing of nuclear weapons on Mururoa and Fangataufa, two remote islands in the Tuamotu group in French Polynesia. Until 1974 the French insisted on conducting atmospheric nuclear tests, resulting in the irradiation of thousands of Polynesians and Melanesians. Wide areas of the South Pacific have been exposed to radioactive fallout by these tests and by the 66 atomic bombs which the Americans dropped on the islands of Bikini and Eniwetok between 1946 and 1958. The Americans have since moved their testing programme to the Nevada desert, but the French have continued to test their bombs in the South Pacific, despite angry protests from virtually every country in the region and in bare-faced contempt of the Rarotonga Declaration.

There are fears that radioactive waste from the French tests, which have been conducted underground since 1974, pose an unacceptable threat to the health and well-being of thousands of Polynesians. Since the French have consistently refused to disclose any data on the amount of fallout their bombs have caused, or on the amount of radiation leaking from their nuclear dump sites, health and environmental risks can only be guessed at.

"Nuclear fallout engendered by the 41 atmospheric tests, made at Mururoa and Fangataufa between 1966 and 1974, is still with us (mostly absorbed in our bodies), and the 63 underground tests made since 1975, instead of diminishing the health hazards, have added several new sources of radioactive pollution," writes Bengt Danielsson, explorer and historian (a member of Thor Heyerdahl's Kon-Tiki expedition in 1947), who lives in Papeete, Tahiti.

French intransigence on the issue of nuclear weapons testing continues to anger the South Pacific states. The issue remains contentious and raw. "No resolution will come until France ceases its mad bombing of the Pacific," notes Danielsson.

Greenpeace: policing the seas

When the *Khian Sea* left Philadelphia carrying 14,000 tonnes of the city's toxic incinerator ash in its holds, the owners, Amalgamated Shipping, thought it would be easy for the ship to dump its cargo in some tropical, cash-short Third World country. They were wrong. In 1988, after 27 months wandering around the world's seas – visiting five continents – the ship was reportedly anchored off Singapore in international waters, having discarded most of the toxic ash somewhere in the Indian Ocean. Earlier, about 4,000 tonnes of the ash were simply left on a Haitian beach, labelled as fertilizer.

The odyssey of the *Khian Sea* was carefully monitored by Greenpeace activists around the world. The ship was tracked across the Caribbean Sea and Indian Ocean and into Southeast Asia and the Pacific. As the *Khian Sea* went from one Pacific island to the next making large cash offers to any government that would take the toxic ash, Greenpeace representatives were busy informing the region's governments about the ship's cargo and what it might mean to their fragile environments if they accepted. In the end, there were no takers.

Largely through the efforts of Greenpeace the ship was unable to find a legal dumping site. "Still, we couldn't track it all the time," admits Sebia Hawkins, Greenpeace's Pacific Campaign Co-Co-ordinator. "We are still trying to find out where the ship dumped its waste ash in the Indian Ocean."

Sebia and her colleagues in the Washington DC office fear that as toxic waste mountains continue to accumulate in the developed world, there will be more voyages like that of the *Khian Sea* to the developing world. "This may be a portent of things to come. We have to be prepared to deal with more of these kinds of incidents."

As the only international NGO which actually polices the seas, Greenpeace is one of the few organizations that attempts to monitor those who treat the oceans as a garbage dump.

Since its inception in the early 1970s, one of Greenpeace's priorities has been a nuclear-free Pacific. In fact, the Nuclear Free Pacific Campaign is the oldest on Greenpeace's agenda. Ever since David McTaggart, one of the founders of Greenpeace, sailed his own yatch into a French nuclear testing zone in Polynesian waters in 1972, drawing world attention to the atmospheric testing of nuclear weapons, Greenpeace has had passionate presence in the South Pacific. In part as a result of the Greenpeace protest and the media coverage of the incident, President Giscard d'Estaing finally abandoned atmospheric testing in 1974. Undeterred by vocal protests from governments in the region, however, France simply went underground with its nuclear programme, just as the Americans and Russians had done in the early 1960s. But there is one significant difference: whereas the United States and Soviet Union now conduct tests within their own home territories, the French continue their tests far away from their home land for the dubious benefits of a nuclear arsenal.

"We now have three major issue areas in the South Pacific," explains Hawkins, "ocean ecology, particularly coral-reef conservation and marine-mammal preservation and management; the production, transport and disposal of toxic wastes; and nuclear weapons testing, both warhead and delivery systems." Within these areas, Hawkins and her colleague Bunny McDiarmid must co-ordinate a campaign that includes the Pacific Rim countries, along with 19 island states and territories. Five of the latter have been targeted for priority attention: Papua New Guinea, the Solomons, the Marshall Islands, Palau, and French Polynesia.

"Dealing with such a plethora of inter-connected issues over such a vast area, we have learned to be sensitive to the development objectives of each island state, including controversial areas such as mariculture, tourism, even the logging of rainforests," explains Hawkins. "The last thing we want is to be labelled 'white eco-imperialists' in this region."

The South Pacific Regional Environment Programme (SPREP), set up with the support of UNEP, is a convenient focus for Greenpeace campaigns. "One of the reasons we like UNEP's regional seas programme," admits Hawkins, "is because they have a mandate from the governments of the South Pacific to do something. The governments want this programme or UNEP wouldn't be here. Greenpeace is playing an active role in both SPREP and the Pacific Forum, lobbying both organizations on specific environmental issues."

Within the framework of SPREP, Hawkins and her crew are working on a protocol which would make all low-lying coral atolls in the South Pacific protected areas where no hazardous activities of any kind would be permitted. Specifically, such a protocol would prohibit nuclear testing along with the storage of toxic wastes and other industrial garbage from the developed world. It might also ban land-based incineration of wastes, because, as Hawkins explains, "small coral atolls and islands are mostly coastal, and any kind of atmospheric or land-based pollution quickly enters the marine environment. On small tropical islands, the environment is all they have."

Another new area for Greenpeace is the question of the management of coastal resources. "We haven't done much on this yet," says Hawkins, "we are still getting our sea-legs on this issue." Greenpeace representatives in the Pacific are already beginning to co-operate with a number of local and regional NGOs, in an effort to develop their capacity to advise South Pacific governments on the efficacy of various management options. "Soon Greenpeace will be in a position to link development/environmental management specialists with local policy people and other NGOs," explains Hawkins.

In a related exercise, the Greenpeace Washington office is involved in the drafting of a Coral Reef Protection Act for the United States. "If we can get something like this through Congress," says Hawkins, "it will have an immediate impact on how the US Trust Territories in the Pacific are managed."

In the end, it is the people of the Pacific who will have to make the difficult decisions. By working together with Pacific-based NGOs, as well as intergovernmental organizations and the United Nations, Greenpeace hopes to help the countries of the Pacific grapple with those choices.

Currently, Greenpeace is expanding its interests well beyond marine concerns. "Since many of the problems afflicting our coasts and seas are land-based, we have to be sensitive to a whole range of issues that might seem at first to be disconnected from the sea," explains Hawkins. Issues such as tropical forest destruction, the mismanagement of uplands, the use of pesticides on croplands, wetland conservation and many others are all being studied for their possible impact on coastal areas.

"I like the diversity of tactics, the various responses to issues that we at Greenpeace have," explains Hawkins. "When we take on an issue, we can go as far as we need to." This militant approach to environmental activism does get results. "I feel very fortunate to have this job, because I know that I am doing everything I can to improve the environment," confesses Hawkins. "But so much remains to be done in the Pacific. If only there was more time."

Biodiversity

As to be expected the Pacific islands are wonders of biodiversity. Geographical isolation has made them that way. Islands that were once connected to continental land masses, such as New Caledonia and Papua New Guinea, are now a refuge for species that were driven to extinction long ago by evolutionary processes on the continents. As a result, most Pacific islands, including coral atolls, have a large number of plants and animals found nowhere else. New Caledonia has many primitive gymnosperms and flowering plants; 80 per cent of them are indigenous. Most of them are also endangered by habitat loss and competition from introduced species.

Isolation alters evolutionary processes in plants and animals. It produces a great variety of species, but they are often unable to cope with invasions of non-native species. So long as islands remain islands, this is no problem. However, once Europeans began colonizing the Pacific in the eighteenth century, they introduced many new, highly-competitive species like cats, rats, dogs and pigs. The new arrivals often annihilated indigenous populations of fauna and flora, which had not evolved ways of dealing with predators and their diseases, or with habitat competition. Wanton destruction of forests and coastal areas killed off others. "Today, there are probably more endangered species per person in the South Pacific than anywhere else in the world," stresses Arthur Dahl, Deputy Director of UNEP's Oceans and Coastal Areas Programme Activity Centre in Nairobi. For instance, there are 54 endangered bird species, or one for every 92,000 inhabitants of Oceania. Equivalent figures show one endangered bird for every 400,000 people in Australia and New Zealand, and one for every 670,000 people in the Caribbean, the other region noted for its extinctions.

So far 15 countries have set aside parks and protected areas to safeguard their natural heritage. In many cases the parks are not much more than proclamations on a piece of paper. Serious efforts must be launched if the South Pacific is to preserve even a small part of its unique communities of plants and animals.

The South Pacific Regional Environmental Programme

The need for a regional approach to environmental problems was recognized early. The South Pacific Conference first voiced concern over the sad state of the region's environment in 1969. Consultations between the South Pacific Commission and UNEP began in 1974. These initial discussions were soon joined by the South Pacific Forum – the political association of independent Pacific countries – through its Secretariat (the South Pacific Bureau of Economic Cooperation) and the Economic and Social Commission for Asia and the Pacific (ESCAP). Protracted discussions led eventually to the decision to organize a conference on the human environment in an effort to create an integrated regional approach to environmental management.

After a long and complicated process, the 22 island states and territories of the South Pacific finally met at Rarotonga in the Cook Islands in March 1982 to endorse the South Pacific Regional Environment Programme Action Plan. Thus, ten years after the Stockholm Conference on the Human Environment launched UNEP, the South Pacific states set up their own programme.

Because existing funding mechanisms were in place to finance the action plan, a trust fund administered by the countries of the region, was not formed. Also, tight budgets meant that no one wanted to be held accountable to pledged donations on a yearly basis. Funds to cover programmes were contributed on a case-by-case basis, with most of the initial money coming from UNEP's Environment Fund.

In the beginning, the programme was handicapped by a general sense of inertia about environmental problems and how to solve them. SPREP's priority programmes suffered from a lack of political commitment on the part of the region's governments. Despite these problems, SPREP was able to carry out a number of important baseline pollution studies and set up a network of scientific institutions and specialists. Since its inception, SPREP has produced over 140 scientific reports and studies. Most countries in the region now have fairly complete resource inventories, along with pollution data, and so can begin to formulate strategies for sustainable development.

In November 1986, 12 independent Pacific states and four other nations with territories in the Pacific adopted the Convention for the Protection of the Natural Resources and the Environment of the South Pacific Region. It had taken nearly five years of hard negotiations, involving more than 20 countries and territories of the region, to reach an accord.

The Convention commits signatories to "prevent, reduce and control pollution in the region from ships, land-based sources, any exploitation or exploration of the seabed, atmospheric discharges, all forms of dumping and the storage of toxic and hazardous wastes." Following the spirit of the Rarotonga Declaration, the states also agreed to prohibit the dumping and storage of radioactive wastes in the Convention area, and to "prevent, reduce and control pollution that might result from nuclear tests in the region". Known as the SPREP Convention, it will enter into force after 10 of the 22 eligible Pacific states ratify it.

With a legal framework, SPREP's action programme can begin to accelerate. But first the programme must develop more diversified and stable sources of finance. Since the preparatory phases of SPREP, over $5 million in contributions have been spent, with more than half coming from UNEP's over-burdened Environment Fund. UNEP is now concentrating its support on specific priority projects, and other parts of the action plan are being funded from other sources.

Obviously, these small island states will never be able to exert much pressure on the international scene. However, by forming a block of interests on certain key issues (like nuclear weapons testing, dumping of radioactive waste, and driftnet fishing) they may still become a persuasive lobby at international forums like the United Nations General Assembly. Their effectiveness, indeed their future, depends on clear policy, decisive action and genuine regional co-operation.

7. The South-east Pacific

The Pacific coast of South America stretches over 10,000 km from Panama in the north to the very tip of Chile. From the dense tropical rainforests of the Colombian coast to the cold deserts of northern Chile it is one of the longest and most varied coastlines in the world.

This region, comprising Panama, Colombia, Ecuador, Peru and Chile, is beset by many of the same problems that afflict the Caribbean and South Pacific. With the exception of Colombia, coastal populations are growing more rapidly than elsewhere. Upland forests have been degraded to make room for subsistence agriculture and grazing lands for cattle and sheep. Mining operations have turned entire mountains into rubble, denuded watersheds and caused massive erosion and siltation of rivers and streams. Industries pump millions of tonnes of pollutants into coastal waters. Rivers which flow into the Pacific often bring in toxic mine tailings and raw sewage. Municipal wastes from coastal towns and cities are dumped untreated into the ocean. Mangroves are being cut down to make room for commercial shrimp ponds.

If some areas have escaped development – like most of Colombia's Pacific coast – it is not because of proper resource management. Usually it is because the area is considered too remote and inhospitable, or too far from traditional population centres and markets.

The economies of the region are still developing, saddled with huge external debt and increasingly forced to export their resources to pay off loans. These five countries have total debt burdens amounting to around $64 billion. Over 25 per cent of the exports of Chile, Colombia, and Ecuador must go to service their debts. With such economics, the South-east Pacific states often find themselves tied to development programmes that sacrifice sustainable resource management for short-term gains. Most of these countries are forced onto the "development treadmill", exporting raw materials and agricultural produce, rather than finished goods. Unfortunately, the prices for basic commodities continue to fall. In virtually every category – metals and minerals, timber, cereals, fruits, beef, and fats and oils – the prices are lower now than in the 1970s. Lower prices result in over-exploitation of limited resources: more marginal land must be cleared to grow more produce or produce more beef; more fertilizers and pesticides must be used to boost yields; more timber must be cut and more minerals extracted. The environmental price for such exploitation is very high.

Recognizing that they have many of these problems in common, the countries of the South-east Pacific, with assistance from UNEP, launched the Action Plan for the Protection of the Marine Environment and Coastal Areas of the South-East Pacific in 1981.

The coastal waters of the South-east Pacific teem with fish and shellfish. Coastal areas also provide some of the best agricultural land in the region. In the mountainous interiors of Ecuador, Peru and Chile, good agricultural land is scarce; in each case the Andes Mountain chain cuts the coast off from the rest of the country. Most of the people living in Peru and Chile are confined to thin coastal strips of land, varying in width from 20–90 km.

Narrow coastal shelves are rich in sea life because the cold and deep Humboldt Current sweeps up the coast of Chile and Peru from the Antarctic, bearing tonnes of bottom nutrients. An offshore version of the Humboldt Current is known as Peru's Oceanic Current. Together these transport systems flow north at the rate of 10–15 million cu m per second; more than the discharge rate of all the world's major rivers combined. This creates nutrient-rich upwellings of cold water along the coast of Chile, Peru and Ecuador. Were it not for this current, the waters of the South-east Pacific might be as barren as much of the South-west Pacific.

Not surprisingly, fishing (and fish processing) is a big industry. Other major economic activities include mining and metal smelting, oil production, food processing, textiles, leather and tanning, forestry, pulp and paper, petroleum refining and chemical production.

Among the diverse habitats of the Pacific coast of South America are large, brown-water river systems. Coastal marshes and estuaries formed at the mouth of these sediment-rich rivers form fecund nurseries for fish and shellfish. Unfortunately, many of them are also polluted with the toxic debris from mining operations and untreated wastes from towns and industries located in the watershed. For example, by the time the Guayas River in Ecuador meets the sea, its waters are laden with industrial wastes, pesticide residues and raw sewage. An estuary once teeming with life is now choked with pollutants. It's an all too familiar pattern. "South America's Pacific coast is the ultimate toilet into which is flushed a multitude of sins," observes Jairo Escobar, UNEP consultant to the Permanent Commission for the South-east Pacific in Bogota.

Cities of the coast

With the sole exception of Colombia, most of the region's people live along the Pacific coast, or within 60 km of it. Roughly thirty-two million people inhabit this coastal zone, most of them crammed into cities and towns. Like the rest of Latin America, the urban population of these five countries is growing more rapidly than the rural. Ecuador's population is growing at the rate of 2.7 per cent a year, while the population of Guayaquil – its largest city – grows at 4.5 per cent a year, doubling in size every 15 years. Most of this growth is attributed to migration from the impoverished countryside. With a current population of 1.7 million, Guayaquil may very well end up with 2.7 million people by the turn of the century. Many of the new arrivals live in slums and squalid shanty-towns built over the remains of mangrove swamps, or next to sewage outfalls and garbage dumps.

Already, 85 per cent of Chile's population is urban, as is 70 per cent of Colombia's and Peru's. In Chile, three-quarters of the people live along a 500-km stretch of coastline between Valparaiso and Concepcion, on 15 per cent of the total land area. The Lima–Callao area of Peru contains nearly 8 million people, 30 per cent of the country's total. Most Panamanians live on the Pacific coast, 760,000 of them in Panama City.

Attempts to encourage the growth of secondary towns away from coastal areas have not had much success. Peru launched a decentralization drive in 1959 in an effort to encourage the development of industrial centres in the largely neglected interior. Despite the creation of industrial parks in cities such as Cuzco, Huancayo and Pasco, the Lima–Callao area still contains 65 per cent of the country's industries.

Colombia, with two coasts to choose from, sensibly chose to develop its more accessible Caribbean side. As a result, Colombia's population is concentrated in the fertile arc extending along the north-central part of the country, and along the Caribbean. Only within the last few years has the government come up with plans to develop its sparsely populated Pacific coast, which at the last census contained only 615,000 people, spread out along 1,300 km of mangrove swamps. Planners in Bogota joke that one of the reasons the Pacific coast has never been developed is because the area's most numerous occupants are mosquitoes, many of which carry malaria and yellow fever. In coastal towns like Tumaco, near the Ecuadorian border, the locals refer to their mosquitoes as "flying hypodermics". Mosquitoes aside, the Pacific coast has been neglected because of a combination of factors: lack of interest on the part of the Colombian government, the absence of infrastructure, and remoteness from traditional markets. Few roads have been cleared through the jungle. Tumaco is connected to the outside world by one potholed road, which is washed out for half the year. The small fishing town of Guapi, some 200 km up the coast from Tumaco, cannot be reached at all except by plane or boat. Only the main Pacific port of Buenaventura has a functional road linking it to the rest of the country. Some on the frontier scoff at the notion of development. One Colombian fisherman from Guapi remarked, "the rivers are our highways, we don't need roads. No one here can drive anything but a boat."

Given the intense level of development along the rest of the Pacific coast, the governments of Ecuador, Peru and Chile wish they had Colombia's options. As more people press upon the coast, pollution mounts and resources are destroyed. "We have heaped this abuse upon ourselves," claims a government planner in Peru. "We have not managed to distribute our growing populations away from the coast. We have not been able to encourage industries to go elsewhere, and so there are fewer jobs in the interior. Our uplands are starving for development, while our coasts are drowning in it."

Coastal pollution

Not only is the South-east Pacific a receptacle for direct discharges of pollutants from coastal cities and industries, it also receives all the pollution washed down from the watershed. Most of the major river systems bring in poisons from mining operations in the Andes, pesticide residues from agricultural land, and tonnes of untreated sewage and municipal wastes from towns and cities.

Nearly every urban area along the coast flushes its wastes into the Pacific untreated. From Porto Armuelle in Panama to the tip of Chile, some 70 major coastal towns and cities deposit everything from raw sewage to solid wastes in shallow coastal waters. Industrial centres add their effluents to the sea.

"There are virtually no sewage treatment plants in Colombia, nor in any of the other countries in the region," points out Jairo Escobar, a UNEP consultant in Bogota. "Industries don't treat wastes either, everything goes into coastal waters – organic wastes from fish and food processing plants and slaughter houses; mine washings; and toxins from tanneries, metal smelters, and chemical plants."

The results can be seen and smelled: coastal waters near every urban and industrial centre are polluted. Panama Bay receives some 34 million tonnes of untreated sewage a year from three rivers and 20 outfalls in Panama City and the port of Balboa. The resultant pollution has created "anoxic" areas in the Bay, where oxygen-starved waters support no fish. Pollution and sedimentation have caused a drastic reduction in the diversity of marine life in the bay.

"Although 95 per cent of Panama City's 760,000 residents have access to potable water, 96 per cent of the city's municipal wastes go directly into the Bay untreated," points out Dr Luis D'Croz, Director of the Center for Marine Science and Limnology at the University of Panama. "We don't have sewage treatment plants and the sewage line outlets go into shallow coastal waters."

To complicate matters, the Bay's sediments also contain traces of Lindane and Eldrin, two toxic pesticides sprayed on croplands along the coast.

Pollution is contributing to a decline in the catches of shrimp and anchovies. The shrimp catch alone is worth around $70 million a year. Trawlers now have to go further out into the Bay in order to find economical quantities of shrimp and fish. As catches decline, over-fishing becomes rampant. Recently, the government imposed a two-month closed season on shrimping, as well as regulations on the type of nets used and the number of boats operating in efforts to bring over-harvesting under control.

Near-shore mussel beds, too contaminated to be harvested, had to be moved elsewhere. Poor artisanal fishermen can no longer make a living from Panama Bay's polluted waters. Instead, some have turned to smuggling.

Meanwhile, the population of Panama City is expected to reach one million by the year 2000. Authorities think the Bay will get worse before some control measures are brought into effect. One solution being discussed is to extend the current sewage outfalls further out into the Bay – a “solution” that will not reduce pollution, but simply spread it out over a larger area.

The only polluted areas on Colombia’s mangrove coast are Buenaventura Bay and Tumaco Cove, both fouled with organic wastes from fish-processing plants, untreated sewage, assorted municipal wastes, and oil. Parts of Buenaventura Bay are now rich in nitrogen and ammonia and poor in oxygen, due to the massive discharge of raw sewage from the city and three rivers. Municipal wastes flow into the Bay at the rate of 54,000 cu m per day. Bacteria (coliform) counts in the Bay are exceptionally high. Floating algal scum, faecal matter and rotting garbage collect around boats in the harbour. A common joke in Bogota is that visitors can smell Buenaventura before they see it. For the people condemned by poverty to live around the Bay, its sorry state is no joke. Many of them have to make a living from what they can gather from the Bay’s waters. It is not surprising that the major cause of death for children under one year of age is diarrhoea, caused by drinking water fouled with human faeces, and eating tainted fish and shellfish.

Ecuador, with its highly developed coastline, suffers from acute deterioration of its coastal waters. Some 20 coastal cities are responsible for pumping 90 million cu m of waste water into the Pacific; 60 per cent of this is accounted for by Ecuador’s largest city, Guayaquil. In addition, Guayaquil Bay receives the wastes from the Guayas River Basin, which covers 34,000 sq km, the second largest on the west coast of South America. Its waters dump sewage, pesticide residues, and heavy metals from mining operations into the Bay. Copper concentrations in the Guayas River and its tributary, the Daule River, exceed safe levels. Some of the toxicity of the copper, however, is thought to be offset by the huge quantities of domestic sewage found in Guayaquil Bay. Health effects from heavy metals have not yet been detected, perhaps because many of the Bay’s poor people suffer from chronic stomach and intestinal disorders brought on by eating seafood contaminated with sewage and drinking dirty water. Viral hepatitis is epidemic, as is malaria.

The waters around the cities of Lima and Callao have turned into cesspools from municipal and industrial effluents. Untreated sewage often washes up on the nearby tourist beaches of Ventanilla, Pampilla and Herradura, prompting city authorities to declare them off-limits for months at a time. The 15,000 industries in the Lima–Callao area dump most of their wastes directly into the sea.

The other source of chronic pollution to Peru's coast is from mining and metal smelting. Mining operations in the Andes routinely flush millions of tonnes of polluted mine washings into rivers and streams, which carry them to the coast. In the mountainous province of Tacna, on the Chilean border, two huge copper mines pump more than 73 million tonnes of mine debris and tailings into local rivers every year. So much of it ends up in the Pacific that near-shore waters contain 21 parts-per-billion of copper, enough to poison a wide variety of marine organisms. In the mountains above Lima, 11 mines deposit 1.8 million tonnes of pollution a year into the Rimac River, which runs through metropolitan Lima before entering coastal waters. Nearly all mine tailings and debris contain a deadly assortment of heavy metals such as copper, cadmium, mercury, lead and nickel.

Further south at Pisco, copper mines pollute the Pisco River with heavy metals, before it flows into shallow coastal waters. Mussels have been contaminated with high levels of copper and cadmium, making them unfit for human consumption. Unfortunately, the people who fish in these polluted waters have little choice.

Untreated sewage is a problem in the Bays of Valparaiso and Concepcion, Chile. The Bio Bio River, which flows into the Gulf of Arauco, just south of Concepcion Bay, is laden with sewage and heavy metals, particularly mercury and lead, from towns and industries in its watershed.

Mining, however, is the major cause of coastal pollution in Chile. In the northern part of the country, copper and gold mines flush nearly 12 million tonnes of polluted tailings and process water directly into the sea. Since mercury is used in refining gold ore, the mercury level in these waters is extremely high. Its effects on marine organisms are not known.

Over-use, or mis-use of toxic pesticides is another source of pollution. In all, the South-east Pacific region uses some 60,000 tonnes of pesticides every year, most of it on croplands. Like the South Pacific region, pesticides banned in the North, such as DDT, are still being applied to crops. Panama Bay, the estuary of the Guayas River in Ecuador, and Concepcion Bay and San Vicente Gulf in Chile are all contaminated with pesticide residues.

"We are currently developing a programme to monitor the use of pesticides in our environment," states Rafael Vasquez Montoya, an officer with the Panamanian National Environment Commission. "The use of really deadly pesticides like DDT is continuing in Panama. We are trying to keep track of what is being used and where."

Degraded resources

With the exception of Colombia, most of the region's tropical forests are being converted into pastureland and cash crops. Panama's Pacific watershed has been logged and replaced by cattle ranches. The country is said to be losing around 36,000 hectares of forest a year. The only region with intact tropical forests is the San Blas, on Panama's Caribbean coast, an area controlled by the Kuna Indians.

Ecuador's coastal mangrove forests are being destroyed to make way for brackish-water shrimp ponds (as described below). Inland, its tropical hardwoods are being cut down and the uplands given over to farms and cattle ranches. Ecuador's pastureland has more than doubled in area since the 1960s, at the direct expense of its tropical forests: the country has been losing around 340,000 hectares a year during the 1980s. There are now only a few ragged remnants of true wilderness left in the entire country. Many of Ecuador's forests consist of secondary growth.

Chile and Peru still have extensive tracts of unlogged woodland in the Andes watershed. Each country has around 30 per cent of its total land area covered by wilderness. But both countries are losing timber faster than it is being replaced; Peru is losing, on average, 270,000 hectares a year, and Chile 50,000 hectares.

There is one factor working in favour of preserving forests. For practical reasons – access to labour and cheap transportation – the many mines operating in both countries tend to be near populated coastal areas and in the Pacific watershed, leaving vast areas of northern and eastern Peru and much of southern Chile free from heavy resource exploitation. So far these sub-tropical and alpine forests have remained inaccessible.

Nevertheless, as upland forests give way to mining, agriculture and ranching, soil erosion is one of the inevitable consequences. In Colombia's central highlands around Bogota, deforested hills are being washed down into the valleys. Hillsides are slit from top to bottom with gulleys carved out by torrential rains. Rivers and reservoirs are silting up.

The Tomine Hydroelectric Dam, 60 km from Bogota, near the village of Sesquile, was faced with closure because of erosion sediment flowing into its 690 million cubic metre storage reservoir. The Tomine Electric Authority decided to fight back with a massive reforestation campaign. Over the past two years, some 1.2 million trees – 80 per cent of them acacia – have been planted on the slopes surrounding the reservoir. The project aims to stabilize 1,200 hectares of badly eroded hillsides by 1991. "Already we have greatly reduced erosion sediment flowing into the lake. And by planting trees, we have reduced the effects of wind erosion as well," states a spokesman for the Electric Authority. So far around 600 hectares have been replanted. Most of the seedlings have been placed on carefully worked up terraces, in an effort to ensure their survival during the rainy season and further stabilize the soil.

In some areas, deep gulleys torn out by heavy rains have been sandbagged in an effort to cut down on the amount of soil reaching the valley and the reservoir. By reforesting bald slopes around the dam, the peasant farmers in the region have also been given another chance. "Before all they could do was watch as their topsoil washed away," observes an official from the Electric Authority. "Now they can see the benefits the trees bring. In some cases they are replanting their own trees."

In this case the watershed has been saved. In many other regions, however, they have been lost. In the uplands of Ecuador, Peru and Chile – particularly in mining areas – deforested hills show huge erosion scars. In some parts of Peru and Chile, entire mountain slopes have broken apart during heavy rains, sending mud avalanches into valleys, smothering villages. Soil erosion in Peru approaches 15 tonnes per hectare per year across the entire country.

Mangrove forests, the regulators of coastal erosion, are being lost throughout the region. All but ten per cent of Panama's mangroves are concentrated along the Pacific coast, where they total around 5,000 sq km. Conversion of mangroves to brackish water shrimp ponds is continuing despite recent economic difficulties. Luis D'Croz calculates that the country is losing one per cent of its mangrove resources every year. "There is some hope, however, that we will still be able to preserve a sizeable portion of our mangroves," explains D'Croz. "With the economy in a tailspin and the costs of opening up new areas getting more expensive, shrimp farming is now less attractive than it was a few years ago."

Colombia's mangroves are largely intact, occupying 280,000 hectares along its Pacific coast – the largest stand in Latin America, outside Brazil. Still, there are fears that as Colombia's Pacific coast becomes more developed, mariculture will accelerate as it has in Panama and Ecuador. Already 5,000 hectares of mangroves have been taken over by shrimp and prawn ponds around Buenaventura.

Ecuador's mangrove resources, except near the border with Colombia, have been decimated and replaced by shrimp farms. Out of a total area of 177,000 hectares, some 60,000 hectares of mangroves have been converted and there are plans to develop another 50,000 hectares.

Since mangroves are essential breeding and nursery areas for shrimp and prawns, cutting down mangroves for shrimp farms simply increases one type of yield at the expense of another. Indeed, commercial shrimp farmers often depend on shrimp fry, hatched in mangrove areas, to restock their ponds.

Coral-reef resources in the region are not extensive. Most Panamanian and Colombian reefs are in the Caribbean, not the Pacific. Because of waters rich in nutrients, fisheries in the region are not reef dependent. The only reefs of any size are off the coast of Panama, and around the Galapagos Islands (Ecuador) and Gorgona Island (Colombia).

Fisheries

The cold nutrient-rich waters swept up the west coast of South America by the Humboldt Current and its offshore counterpart ensure that the waters off the coasts of Chile and Peru are teeming with marine life. Local upwellings in the coastal waters of Ecuador, Colombia and Panama are met by wedges of freshwater brought in by tropical rivers which bring nutrients – and pollutants – from the interior.

The annual take of fish and shellfish from the region amounts to just under 10 million tonnes, or roughly fifteen per cent of the world's total commercial harvest from the sea. Peru and Chile account for all but a fraction of that tonnage. In 1985 their deep water fleets caught over 9 million tonnes of fish, mostly anchovies, sardines, hake, and jurel. In addition, Peru produced 1.4 million tonnes of cultivated crustaceans, while Chile cultured 7,000 tonnes of clams, oysters and algae. Most cultured shellfish are exported to Japan and North America.

Panama Bay is rich in marine life because of an upwelling of deeper, nutrient-laden water. Nearly ninety per cent of the Panamanian commercial fishery is based in Panama Bay and Panama Gulf. Three species of white shrimp are caught, amounting to 7,000 tonnes a year. Many anchovies are also taken, and scallops are harvested by artisanal fishermen working in the less polluted southern part of the Bay. But mounting pollution is exacting a toll on the Bay's fisheries. "At the moment, we are not getting the maximum sustainable yield from our fisheries because of over-fishing and pollution of near-shore waters," states D'Croz.

In Ecuador the fishing industry ranks second, after oil production, as a source of foreign revenue. The total amount of fish and shellfish exported amounts to around 300,000 tonnes a year. The export of farmed shrimp is also accelerating, as is organic pollution from shrimp farms, which is pumped into coastal waters.

By contrast, the Pacific coast of Colombia is unique: it is virtually undeveloped. From the metropolis of Buenaventura in the north to the small fishing village of Tumaco near the Ecuadorian border, the mangrove coast stretches for over 600 km. It is one of the most unspoiled mangrove wildernesses left on earth, broken only by meandering brown-water rivers that flow lazily into the Pacific. There are no roads through the jungle.

From the air, the sight of so much unbroken tropical greenery is breath-taking. Along the entire length of the mangrove coast only one town of any size appears on the map – Guapi, situated at the mouth of the Guapi River. The few fishing settlements consist of tin-roofed wooden shacks, bounded on all sides by imposing mangrove swamps. During the last century the sugar boom prompted the establishment of a few plantations, worked by slaves from West Africa. But the plantations failed, unable to withstand the effects of disease and their remote location.

Unable to make a living from the land, the people of the mangrove coast have turned towards the wealth of the sea. Everything from giant prawns to grey whales lives here, but it is the langustino shrimp (*Penaeus occidentalis*) which sustains most of the estimated 5,000 people who ply the coast in simple dug-outs or motor-boats. Since there are no large trawlers operating in these waters, the region's fisheries are under-developed. Only around 17,000 tonnes of fish and shellfish are taken every year; a mere fraction of the annual potential yield which has been calculated at between 130,000 and 156,000 tonnes.

El Niño

The disruptive, and destructive, weather patterns which periodically sweep through the Pacific have come to be known collectively as El Niño, The Child. However, its effects are anything but child-like.

When this bizarre weather pattern hit the west coast of South America in 1972, the anchovy fishery, worth millions of dollars, collapsed. Years of over-harvesting had of course, contributed to its demise.

El Niño can perhaps best be understood as a sort of ocean drought. Climate controlling ocean–atmospheric interactions are disrupted on a large scale. Temperatures climb and surface waters become much warmer than normal. Warmer surface temperatures cause more nutrients to be sedimented out of the water column, depleting coastal waters of the food needed to sustain marine life, including commercially important fish. With food sources reduced and water temperatures becoming unbearably warm for cold water fish, they go elsewhere.

Those creatures who cannot leave, learn to take advantage of the warmer waters or perish. Hence, during the last devastating visit by El Niño, in 1982–3, coral reef communities in the Gulf of Panama simply died off as water temperatures soared to 31° Centigrade. Tropical crabs, on the other hand, were able during the same period to expand their breeding ranges further south thanks to the warmer waters.

The El Niño of 1982–3 turned out to be the worst of this century. Heavy rains drenched Colombia, Ecuador and Peru. In northern Peru, rainfall was 340 times greater than normal. Some rivers carried over 1,000 times their normal flow. The widespread flooding that ensued took a terrible toll on crops, livestock, roads, bridges, schools and homes. In Ecuador, 40,000 families were made homeless by flood waters. Unable to catch anything in the warmer surface waters, tens of thousands of fishermen across the Eastern and South Pacific were left idle. In Southeast Asia and Australia, prolonged droughts ruined crops and turned soils to dust.

"We now have a regional research programme in place to study this phenomenon," says Jairo Escobar, "it is a really big problem for us. The last time it struck, there were droughts across the Pacific island states and Southeast Asia, while Colombia, Ecuador, and northern Peru suffered heavy rains. Our entire economy was affected."

The South-east Pacific Action Plan

The Action Plan for this region is nearly a decade old. The five countries of the South-east Pacific gathered in Lima, in November 1981, to approve the "Convention for the Protection of the Marine Environment and Coastal Areas of the South-East Pacific". At the same time they also approved the "Agreement for Regional Cooperation to Combat Pollution of the South-East Pacific due to Oil Hydrocarbons and other Noxious Substances in Cases of Emergency". The Permanent Commission for the South-east Pacific (CPPS) was designated as the co-ordinating agency for the Action Plan. The Lima meeting was the product of four years of preparatory work carried out by the Permanent Commission for the South-east Pacific and the region's governments, in co-operation with UNEP.

Subsequently, in 1983, another important protocol was added to the Convention: the "Protocol for the Protection of the South East Pacific Against Pollution from Land-based Sources". They were all duly ratified and came into force in 1987.

One of the first things that the CPPS did was to set up a network of national contact points in each of the participating countries. Then a research programme was worked out and approved in an effort to assess the state of the marine environment and to pinpoint polluted areas in need of priority attention.

"When we started this programme, we had no information at all as to the state of our coastal waters and marine resources," points out Jairo Escobar. "We are still trying to upgrade our database, but at least a good beginning has been made."

The pollution research and monitoring programme now has its own network and its own name: CONPACSE. So far, baseline pollution studies and assessments have generated a wealth of data and scientific expertise – some forty technical reports are now available to national authorities trying to grapple with coastal pollution. CONPACSE research has involved 42 regional institutions.

In addition to this research work, CPPS has supervised the training of 320 experts in marine pollution evaluation, environmental impact assessment, and analytical techniques. "We have two phases to the programme," states Escobar, "to provide basic information on the state of the region's resources, and to promote sustainable land-use planning."

As data continue to be gathered, the five countries of the region are now in a position to begin formulating, together, land-use management strategies that will have a direct impact on reducing coastal pollution.

A priority Action Plan has also been worked out for the region. It includes three areas: the control and treatment of municipal and industrial wastes; the development of artisanal fisheries; and the promotion of sustainable mariculture and aquaculture industries.

"We have managed to educate the experts," notes Escobar, "next we have to educate the public about the need to conserve coastal resources and to manage them properly. We have already launched a programme in Buenaventura to teach high school students about the marine environment and its importance to them."

At the same time the commission has established links with its counterpart in the South-west Pacific, in order to facilitate co-operation on important pan-Pacific issues, such as efforts to make the entire South Pacific a nuclear-free zone. With a Trust Fund set up and money flowing in, Escobar is confident about the future. "We may have a lot of problems, but we now have the information we need to begin to solve them."

8. East Asia

Virtually all east Asia's major cities are coastal, as are countless thousands of villages. It is here (and in south Asia) where sheer numbers are overwhelming coastal resources. More people than ever before are now dependent upon coastal areas for their livelihoods: for fishing, mariculture, forestry, building materials, agriculture and tourism. Yet the very resources they depend on for survival are being needlessly over-exploited and destroyed.

Mangrove forests are cut down to make room for urban expansion, rice paddies, coconut plantations, and brackish water fish and shrimp ponds. Coral reefs are smothered by siltation brought in by coastal dredging, mining development and deforestation. Offshore reefs are blasted apart by dynamite fishermen, heedless of the destruction they inflict on coral communities. Wastes from industries and municipalities are simply dumped into the sea untreated. Rivers bring in more pollutants: sewage, pesticide residues and industrial effluents.

The economies of Thailand, Malaysia, Singapore, Indonesia and the Philippines vary a great deal. Thailand and Singapore are booming, Malaysia is muddling along, and Indonesia and the Philippines are in crisis. Without coastal management plans in place the brunt of unplanned development and economic expansion will be borne by the coasts. Without sustainable management of resources, coastal economies will not be sustainable either. Coral reefs are pillaged by coral collectors and dynamite fishermen, ruining tourism. Mangroves are cut down for fish farms, reducing offshore shrimp and fish catches. In too many places the link is not made between what happens on land and what happens in the sea.

Like the countries of the South-east Pacific, Southeast Asian nations are caught in a trap of debt and development. As debts rise, more resources must be exploited to pay the interest on loans. The Philippines is poorer now than it was 20 years ago; nearly sixty per cent of the population lives below the official poverty line, which has been set at $120 a month. Fiscal mismanagement is blamed, but international debt is also a factor. In 1988, the Philippines was condemned to see 28 per cent of its export income swallowed to service debts.

A sea of islands

Nearly all the territory of Indonesia and the Philippines – the two biggest island archipelagos in the world – is coastal. Malaysia, a thin peninsula, is bounded by sea. Singapore is an island. Only Thailand has some land-mass to spare.

South-east Asia's seas are dotted with islands. Indonesia's 5,000-km long archipelago contains nearly 14,000 islands; the Philippines has 7,100; and there are hundreds in the Andaman and South China Seas. Disregarding Canada and Greenland, Indonesia and the Philippines also have the longest coastlines in the world. Indonesia's serpentine coastline exceeds 54,000 km. If stretched out in a straight line, it would encircle the earth twice.

The geography of South-east Asia is a geography of coasts, more so than perhaps any other region. Many of Indonesia's islands consist entirely of coastal zones. With the exception of the largest islands, like Borneo, what happens inland has coastal repercussions. In the Philippines, which lost over 60 per cent of its tropical forests during the past four decades, increased erosion of the uplands has translated into massive sedimentation of shallow coastal waters.

The location of these island and peninsular states has forged the region into one of the world's major trading crossroads. Indonesia, Malaysia, the Philippines, Singapore, and Thailand form one biogeographical unit, separating Asia from Australia and the Pacific from the Indian Ocean. The region covers almost 9 million sq km, or 2.5 per cent of the world's ocean surface.

Travel writers often refer to this area as the "Asian Mediterranean". Weather may have something to do with the comparison. Both the Mediterranean and South-east Asia have definite wet and dry seasons, although Asia's are governed by the monsoons. Both areas also concentrate people – and pollution – along their varied coastlines. Beyond that the similarity ends. The Mediterranean has a relatively stable average depth of around 1,500 metres, while in Southeast Asia there are extensive shallow continental shelves and deep basins perforated with trenches and troughs. The whole area is studded with volcanic and coral islands.

Seasonal trade winds allowed the "spice islands" to be visited by Arabs from the west, Chinese traders from the east and Europeans from both directions. While trying to find a short-cut to the East Indies – meaning the Indonesian and Philippine archipelagos – Colombus accidently discovered North America. Not knowing where he was, Colombus called the islands of the Caribbean the "Indies" and their inhabitants "Indians". Meanwhile, the real "East Indians" quickly became major world suppliers of spices, rice, coffee, tea, sugar, palm oil, rubber and tin.

The area has been frequently colonized. The Philippines were Spanish, and then American, while Indonesia remained Dutch until 1949, when it gained independence. Malaysia and Singapore were English. Only Thailand was never under foreign domination.

Today the region finds itself on a major oil route. All Middle East crude destined for Japan, China, Hong Kong and South Korea must pass through either the narrow Strait of Malacca or the Lombok Channel. One of the reasons for the US military bases in the Philippines is to protect these vital shipping lanes.

Most of the region's people are coastal, or live within 50 kilometres of the sea. There is a wide assortment of racial and religious groupings: anthropologists claim there are at least 110 different cultural and racial groups in the Philippines alone, speaking some 70 different languages, ranging from the Muslim Malays in the southern Sulu Islands to Episcopalian Igorots in the mountains of Luzon. Indonesians speak 250 languages and dialects, making communication between villages, not to mention islands, somewhat of a challenge. Thailand contains more than 30 distinct ethnic groups.

The one thing most South-east Asians have in common is a dependency on the sea for food, livelihood, and transport. The Buginese of Indonesia, known as the "sea gypsies", spend most of their lives on the sea. "They are born, grow up, get married, have children and eventually die on the sea", says Jakarta journalist Winnarta Adusibrata.

Like the rest of Asia, the populations of these five countries – with the exception of Singapore – continue to grow, but not nearly as fast as during the 1960s and 1970s. Both Thailand and Indonesia launched highly successful family planning programmes in the 1970s. Only Malaysian and Filipino populations continue to expand at more than two per cent a year. Indonesia initiated its two-child family campaign ("Dua anak cukup") in 1970 with assistance from the United Nations Population Fund. Today, Indonesia's large population is growing at an acceptable 1.8 percent. Still, this national figure conceals some troublesome trends. The growth rate of Javan coastal cities like Jakarta and Surabaya approach four per cent a year, doubling their populations every 17 years. Swelled by an influx of migrants from the uplands, the majority end up in squatter settlements or slums, under-employed or under-paid.

Emil Salim, Indonesia's Minister of State for Population and the Environment, has a job many consider impossible. Not only must he keep an eye on the country's population growth-rates, but he must also try to balance the needs of increasing human numbers against dwindling resources. As a cabinet minister, Salim must hold his own against powerful colleagues who want development, no matter what the environmental costs. "He must walk a tightrope every day," remarks an Indonesian official. "It is not an easy place to be. In order for him to be effective, he has to have the support of the President."

Today, Salim's ministry plays only a co-ordinating role in efforts to stall population growth. The real authority lies with the National Family Planning Co-ordinating Board (BKKBN), which operates with a massive budget and support from the United Nations Population Fund. Despite the fact that Indonesia's population is still growing at the rate of three million a year, family planning and widespread access to contraceptives and basic health care have made a big difference. The fifth most populous country in the world "would be in infinitely worse shape were it not for family planning and the two child family campaign," points out Erna Witoelar, Director of the Indonesian Consumers' Organization.

Salim's most difficult task is to try to balance environmental protection needs with national development goals. Contradictions abound. As more forests are cut for timber export and to expand agricultural production in the uplands, hydropower reservoirs silt up, reducing the lifespan of hydro-electric dams to a mere 30 years. Rivers, swollen with silt, dump more of it into coastal waters, killing mangrove forests and coral reefs. Conserving some timber in vital watersheds would provide a little insurance against soil erosion in the hills and silted-up dams and rivers. But in the rush for development, such considerations are often overlooked. With less tree cover in the hills, droughts become more severe and floods more frequent.

In his efforts to make the development of his country sustainable, Salim continues to struggle with the harsh realities of an international marketplace that seems to penalize those countries forced to sell basic commodities. "My greatest enemy in terms of environment protection is world trade," observes Salim. "Today, we have to export three times as much timber, for example, to buy one tractor as we did in the 1970s." Saddled with a national debt of $50 billion, and a devalued dollar, Indonesia must export 40 per cent more of its resources – timber, oil, gas, rubber and bauxite – in order to meet interest payments on its loans. The environmental price for this is almost extortive. "We can't just say, hands off the tropical forests," argues Salim. "We do what we can to conserve resources, but in the end we have to export in order to develop."

As agricultural land shrinks on a *per capita* basis, coastlines are becoming inundated with people. Hence, coastal management has become one of Salim's priorities and one of his biggest challenges for the decade ahead. "We still must press ahead with family planning programmes, the re-distribution of our population from over-crowded islands like Java, and the intensification of agricultural production," claims Salim. "But at the same time, coastal area management has become vital for our survival."

During 1989, Salim introduced a sustainable development strategy into national development plans for the next five years. One of the main features of this plan is its attempt to manage coastal resources. "Our challenge is to find ways to incorporate environmental considerations into development plans," says Salim. "We must find ways to translate this message into language that planners and economists can understand."

Resource destruction

So far, South-east Asia's record of resource management is disappointing. Part of the reason is the overwhelming numbers of poor people forced by circumstance into abusing and over-using common resources. Another reason is government's inability, or unwillingness, to come up with realistic management plans for utilizing resources, particularly coastal resources, in a sustainable fashion. Coastal area management schemes have been drawn up by government planners and high-priced consultants, but few get implemented outside of test sites. Furthermore, so long as most of these schemes ignore the people they are meant to benefit, little progress will be made.

Mangrove destruction in South-east Asia is the most extensive anywhere. Although the region still has about five million hectares left, this figure is thought to represent less than half of the original area. Between 1920 and 1988, the mangrove area of the Philippines was reduced from 500,000 hectares to a mere 38,000 hectares. Much of the destruction took place over the past two decades; between 1967 and 1975 the country lost 24,000 hectares of mangrove forests every year. Most of them were clear-cut for their timber, exploited for tannin (a natural preservative), or were cut down and converted into brackish water ponds for the cultivation of prawns and milkfish. Others succumbed to urban expansion or were filled in to create rice paddies.

On the island of Negros, in the southern part of the Philippine archipelago, only one significant stand of mangroves survives on the east coast, in Bais Bay. Even these consist mostly of secondary growth. Nevertheless, a portion of Bais Bay's mangroves have been declared a sanctuary and are now the subject of intensive study by marine biologists from Silliman University in Dumaguete City. With the loss of the island's mangroves (along with most of its coral reefs), fish and shellfish stocks declined dramatically. The poor fishing communities on Negros found it increasingly difficult to make a living from the sea. Competition for limited numbers of edible fish encouraged destructive fishing practices such as dynamite fishing and the use of poisons, killing fish indiscriminately and further damaging valuable marine habitats. As on many other islands, the pattern of destruction worsened as desperate fishermen tried to feed their hungry families by any means possible. Unlike many other islands, however, the fishing communities around Bais Bay began to rebuild their ravaged resources by planting new mangroves and by constructing artificial reefs and fish shelters.

Ricardo lives in the small fishing village of Okiot on the north rim of Bais Bay. Most of the fishermen in this community trap fish in "corrals" made from bamboo and nylon mesh. A corral is simply a fence in the water with an opening at one end. Fish enter the corrals during the day while feeding, sometimes attracted by fish shelters placed inside the enclosure. At night, fishermen using lights attract the fish in the corral to an opening where they are caught with dip nets. "Mostly we catch rabbit fish this way," says Ricardo. "And during the day women pick over the tidal flats for abalone, oysters, mussels and crabs."

Traditionally, Ricardo's village has always trapped fish, rather than raising them in pond cultures. But with fish catches remaining low, Ricardo and his neighbours are thinking of building brackish water ponds to raise milkfish from fingerlings. One of the reasons the community must begin to farm fish is because both their mangrove and coral reef resources have been severely damaged. "We have to watch our remaining mangrove forests very carefully," points out Ricardo, "otherwise people from other islands and sometimes farmers from the hills come to cut the trees at night." Since rabbit fish use mangroves for part of their life cycles, the people of Okiot see the trees as a direct link in their own food chain.

Like the fishing communities further north, Okiot plans to reforest the coast with mangroves in an effort to restore fish stocks. They have even formed a sea-patrol which cruises the shallow waters of Bais Bay apprehending fishermen who use dynamite and poison – instead of hooks and lines, or nets – for catching fish.

The resource problems of Bais Bay are not unique to the Philippines nor to South-east Asia in general. As fish stocks continue to decline, many people on Negros and other islands have been forced to become small-scale fish farmers. At the same time entrepreneurs clear mangroves to build brackish water ponds for raising tiger prawns or shrimp for the export market. But as more ponds are built, there is less room to restore mangrove forests and less incentive to protect coral reefs. Ricardo's village is trying a middle course. The fishermen of Okiot are conserving their mangroves and farming fish.

Unfortunately, the steps taken on the east coast of Negros are rarely followed. Across South-east Asia and the Pacific, familiar patterns of destruction are repeated. By 1970, over 1,200 sq km of mangrove forests on Sabah (Malaysia) – 40 per cent of the total – were allotted for woodchip export to Japan. Today, Malaysia is losing 5,000 hectares of mangroves a year for the production of woodchips. Thailand has lost a full 20 per cent of its mangrove forests over the last ten years, cut down for timber and to make room for fish ponds. Most of Singapore's mangroves have been removed by urban expansion.

Perhaps the greatest destruction of mangroves is taking place in Indonesia. Currently, more than 2,000 sq km of mangroves are being exploited by the Japanese woodchip industry. In addition, some 10,000 sq km of mangroves have been converted to brackish water ponds, called *tambaks*, in order to cultivate prawns, shrimp and milkfish. On the east coast of Sumatra, over 32,000 hectares of mangroves have been cleared for *tambaks*. Nearly the entire north coast of Java, once lined with mangroves, is now lined with *tambaks*. At Cilacap, on the south coast of Java, mangroves have been clear-cut to build fish ponds, and at least 50 sq km have been turned into agricultural land to produce rice, coconuts and other cash crops.

In the province of Kalimantan on the large island of Borneo, extensive tracts of mangroves have been cleared to make room for new coastal settlements – part of Indonesia's controversial transmigration programme which relocates people from the overcrowded islands of Java, Madura, Bali and Lombok. So far, some 700,000 Indonesian families have been sent to these new lands.

Indonesia's loss of mangroves has affected the livelihoods of millions of small-scale commercial and artisanal fishermen. Offshore shrimp and prawn catches have fallen, as have landings of groupers, jacks, mullet, and snappers.

Declining fisheries is also linked to the loss of coral reefs. Thirty per cent of the world's coral reefs, representing all known genera of corals and all morphological types of reefs, are found in South-east Asian seas, but they are being degraded and destroyed at a shocking rate. When the Philippines carried out a survey of 632 reefs in 1982, researchers discovered that two-thirds of them were in poor to fair condition, with less than 50 per cent live coral cover. Since then, the reefs have deteriorated further – today less than ten per cent have healthy, undamaged coral. The widespread destruction is attributed to siltation from massive deforestation, uncontrolled blast fishing and use of poisons, and coral mining for the building industry and for export.

While conducting the 1982 coral reef survey, researchers from Silliman University discovered that five coral islands off the coast of Cebu had actually disappeared. Victim of the rapacious mining interests on Cebu, the coral had been excavated, piece by piece, for use in making cement and building materials. "The main terminal building at Cebu City airport is made partly from coral tiles," states Professor Angel Alcala, head of the Marine Laboratory of Silliman University in Dumaguete City on Negros. "The tiles look like marble, but they are really made from coral." A national ban on the export of coral products has had little effect: too many people find the profits irresistible, and there is no enforcement of the ban.

Indonesian reefs are under increasing stress from both commercial and artisanal fishermen. Thailand's corals are suffering from siltation churned up by huge bucket dredges that scour the bottom sediments of near-shore waters for tin. It can take weeks for the suspended sediments to drift back to the bottom. Often they end up smothering nearby seagrass communities and coral reefs.

Conservation efforts have not yet been able to preserve much of the diversity of the region. But that may change as governments put more emphasis on conserving genetic resources. More parks and protected areas are being established. Indonesia has set aside 17 parks and reserves with marine areas. In the early 1980s, government policy called for the establishment of 10 million hectares of marine and littoral protected-areas by 1990. A Marine Conservation Data Atlas was produced with assistance from IUCN, to help in planning a full-scale system of marine parks. Unfortunately, too many of the declared sites are parks only on paper, without a management programme to guide their operation.

Coastal pollution

Parts of South-east Asia's coasts are among the most polluted of any region. With 210 million people living on or near a coast, many in expanding cities like Bangkok, Jakarta and Manila, heavy pollution is inevitable. With the exception of Singapore, none of the cities or towns of the region has adequate sewage treatment plants. Open drainage canals channel raw municipal effluents and industrial wastes directly into coastal waters, or into rivers which transport their wastes to the sea.

Nearly all of South-east Asia's river systems are polluted with organic wastes, sewage, plastics, and junk. Rivers like the Juru (Malaysia), the Pasig (Philippines), and the Chao Phraya (Thailand) are open sewers by the time they reach the sea. All of Metro Manila's rivers are so full of pollutants that they are said to be biologically dead. Because of the amount of municipal and industrial wastes dumped into the Bays of Manila and Jakarta, large sections of each suffer from a loss of oxygen and cannot support much marine life. The Chao Phraya estuary in the Gulf of Thailand is so polluted from Bangkok's wastes that its murky brown waters can no longer sustain shellfish like clams and oysters. Artisanal fishing for shrimp and prawns has stopped. Those lacking motor boats have had to abandon fishing altogether; they are unable to catch anything worth eating in the Gulf's near-shore waters.

The raw wastes of 8.5 million people and 10,000 industries end up in Manila Bay. Its polluted, malodorous waters are filled with human faeces and rotting garbage. "The bacteria (coliform) count in Manila Bay is over 100 times above safe standards," admits Celso Roque, Under Secretary of the newly-formed Department of Environment and Natural Resources in Manila. "Our first priority is to reduce the domestic sewage loads to Manila's five rivers. Industries have been asked to clean up their own mess." Manila Bay is also plagued by high levels of mercury, cadmium and lead in its sediments.

Unfortunately, a coastal-zone management plan was turned down by the government on the grounds that it would interfere with decentralization efforts. Meanwhile, the entire waterfront of Manila is covered with squatter settlements: flimsy bamboo huts built on waterlogged rafts, half-sunken barges, or perched on wobbly stilts over fetid waters. More than two million people in Metro Manila are crammed into these floating slums. Others occupy any unused land they can find, such as the empty field surrounding the unfinished National Insurance Building. One entire community makes its living from Manila's huge garbage dump, called "Smokey Mountain", recycling whatever they can salvage.

There are many small fishing communities around Manila Bay, despite its sorry state. They have nowhere else to go. Children bathe and swim in its filthy waters. Fishermen must go further and further to catch fish. Those who cannot afford the extra petrol for their motors must somehow make a living closer to shore. The Bay's waters are so polluted that freshly-harvested oysters smell as if they've been dead for a week: clam and oyster farms in the southern part of the Bay should be moved elsewhere for health reasons. City officials agree, but ask "where to?" Meanwhile, the polluted products from the Bay are sold to the city's poor. Stomach and intestinal diseases are reaching epidemic proportions. Many children under one year of age die from chronic diarrhoea.

The fishing communities on Jakarta Bay, just north of the capital of Indonesia on the island of Java, pay a similar price for their poverty. Over the past 20 years the Bay – once rich in fish and shellfish – has turned into an open cesspool. Millions of tonnes of industrial and municipal wastewaters are discharged into the Bay every year without treatment of any kind. One village struggling against the effects of pollution and unplanned urban growth is the fishing community of Muara Angke.

"I've spent nearly my entire life on the sea," says 78-year-old Sho Boen Seng, the unofficial head of Muara Angke. "Twenty years ago fish were plentiful in Jakarta Bay, but not now. Today our fishermen must go out more than 12 miles from the shore to catch quantities of fish," explains Seng. The round-trip, in small out-board boats, takes at least six hours. Even the bigger boats have to stay out much longer to land profitable catches of fish than was the case in the 1950s and 1960s.

Pollution and sedimentation are the main culprits. Coliform (bacteria) counts and toxic metals in the Bay exceed US Environmental Protection Agency (EPA) standards on a regular basis. Parts of the Bay are becoming starved of oxygen due to the heavy input of nutrients from untreated sewage and industrial wastes. Sedimentation from dredging and landfilling operations have helped to obliterate vital near-shore spawning areas for shellfish like shrimp and clams.

Pollution has reached such proportions that much of what is caught in the Bay is contaminated and unfit for human consumption. Unfortunately, the poor fishing families of Muara Angke have little choice: they must eat what they catch in the Bay's waters. Most people cannot afford the bigger boats needed to fish in open seas. The poorest are condemned to fish with simple dip nets off oil-covered jetties in the harbour. As a result, many residents suffer from chronic intestinal and stomach disorders likey dysentery, diarrhoea, and gastro-enteritis.

Muara Angke, built on land reclaimed from the sea, became a special fishing village in 1977. The government-sponsored village was put up as a place for subsistence and small-scale commercial fishing families who were being displaced by urban sprawl. It still lacks many basic amenities, such as potable water and sanitation. The village's own wells have become too saline, so fresh water is brought in by vendors, who sell it door-to-door by the bucket. Sanitation facilities are also primitive, consisting mostly of pit latrines or open drainage systems which channel sewage and household slops directly into the Bay.

Like all of Muara Angke's 3,000 residents, Seng and his wife and daughter live in a simple house made from breeze blocks and concrete with a tin roof. Seng's main work consists of trying to get the government to pay more attention to the needs of the village and its people. In 1980 the government banned trawlers from Jakarta Bay in an effort to give small-scale fishermen a better chance to make a living. It was a good gesture, but came too late to have much impact on fish catches. The Bay's shallow waters were already so degraded and over-fished that trawlers would have left anyway.

Meanwhile, Muara Angke remains caught between the old ways and the need to adapt to change. "The government provided small loans to local fishermen so they could buy cheap out-board motors for their boats," explains Seng. "They even trained us to use and service them. But too many fishermen here still fish with primitive gear. And our small harbour still lacks cold-storage facilities." The fish that are brought into Muara Angke have to be sold the same day in the local fish-market, or salted and dried in the sun.

The low prices paid to the fishermen of Muara Angke for their fish are also determined more by middle-men, the money-lenders who control finance in the village, than by open market forces.

Seng, who has the equivalent of a high school education, thinks many of the village's problems come from a lack of education: "Most of the fishermen here have no more than a primary school education; and many of the children don't attend school because they must go to sea at an early age to help support the family." The other big problem is the inability to communicate with each other. There is no common language in Muara Angke. Many fishermen don't even understand Indonesian. Instead, they speak a variety of local dialects.

In an effort to forge a real community, Seng is striving to organize the fishermen into a co-operative union. This way they would be able to get better prices for their fish, and they could draw on co-operative reserves during lean times. "The problem is most of these poor, uneducated fishermen don't want to put in their own money," explains Seng. "They think they'll never see it again. But without a co-operative, this village will always be at the mercy of the money-lenders. We will never be able to stand on our own feet."

Indonesian coastal pollution is not limited to Jakarta Bay. Since over 75 per cent of the country's cities are situated along the country's 54,716 km of coastline, many coastal areas are inundated with untreated wastes. High coliform bacteria counts are common in Ambon Bay and the Flores Sea. The entire coastline of Java is crowded with cities and towns, interspersed with shrimp and fish ponds and agricultural lands. The wastes from cities, shrimp cultures, milkfish ponds and industries end up being dumped in the sea, creating oxygen-starved conditions in shallow coastal waters.

In Thailand, many of Bangkok's slums are water-bound. Thousands live on water-logged barges moored on the city's many canals, or along its four major rivers: the Chao Phraya, the Thachin, the Maeklong and the Bangpakong. These four rivers deposit roughly 10,000 tonnes of raw sewage and municipal wastes into the Gulf of Thailand every day.

Although Bangkok has drawn up water-quality guidelines and has initiated a Master Sewerage Plan for the city, half the population would still not be covered by a sewage treatment network by the plan. The World Bank has calculated that the cost of connecting all of Bangkok's residents to sewage lines would be a prohibitive $1.4 billion. An alternative system utilizing the city's many refuse-clogged canals is now being developed.

Along Thailand's 2,600 km of coastline, some twenty major urban centres and 9,000 fishing villages can be found. Over twenty million people live around the Gulf of Thailand alone. Tourist centres are mushrooming, particularly in the resort area of Phuket, in the far south. The household and municipal wastes from all these communities are flushed into coastal waters.

Industrial pollution often has immediate and disastrous effects on coastal ecosystems and on the economies of villagers dependent on the sea for their livelihoods. The small fishing village of Dauis on Bais Bay on the island of Negros, the Philippines, has had a sugar mill for four decades. The mill regularly pumps untreated toxic wastes from sugar processing, including bagasse, into the Bay's shallow waters. If the Bay is polluted with sugar wastes at the time of the year when young crabs and shrimp fry are developing, it can eliminate an entire half-year's income for the village.

In July 1988, one such discharge killed off around 200,000 blue crabs, along with thousands of shrimp, which the villagers of Dauis could have harvested at the end of the year for a sizeable income. Masses of fish were also found dead and dying in the Bay. The people of Dauis estimate they have lost some 20 million pesos a year since 1979 because of pollution from the sugar mill and a paper mill further up the coast.

"We have been protesting to Manila since 1954," says one angry villager. "The people who led the original protests have all died, but the pollution remains."

Around 10,000 people in the region, including eight fishing villages, depend on Bais Bay for their survival. The sugar mill employs only a few technicians, and it is losing money.

Fisheries

In order to avoid the tragedy of Dauis, many coastal fishing villages are turning to shrimp and fish ponds, where inputs can be controlled. The problem is that many mariculture and aquaculture farms are built on the remains of mangrove swamps, taking away critical habitats for a wide variety of fish and shellfish. As a result, wild stocks are reduced. This creates what one marine biologist terms "the Mobius Loop of resource degradation". As wild stocks get more difficult to find and catch, more people are encouraged to build their own fish or shrimp ponds. More mangroves are cleared. The process repeats itself again and again until there are no more mangroves or lowland coastal forests.

In some areas, as on the north coast of Java, mangrove destruction can touch off a chain reaction of coastal degradation. Marta Vannucci, Director of the UNDP/UNESCO Regional Mangrove Project, explains:

> The most senseless exploitation is the clear felling of mangrove areas to build shrimp ponds that still rely on the recruitment of larvae from the reproductive stock at sea for seasonal re-stocking. Likewise, shrimp ponds built on clear-felled mangrove land cease to be economically viable enterprises after a couple of years because natural recruitment of larvae is no longer possible. New mangrove areas are cleared to set up new shrimp ponds and further destruction follows. Thus, a downward spiral of degradation from a more to a less productive system takes place.

There is a clear need for regional management of fisheries. "Nearly all Asian waters within 15 kilometres of land are considered over-fished," explains Ed Gomez, Director of the Marine Science Institute at the University of the Philippines in Manila.

Since South-east Asians get most of their protein from the sea, fishing and various mariculture operations have an added importance. Filipinos derive at least 60 per cent of their animal protein from fish, Indonesians 63 per cent. "You have to keep in mind that the average Asian eats rice and fish with nearly every meal," points out Dr John McManus, also of Manila's Marine Science Institute.

Moreover, millions of artisanal and commercial fishermen make their living from the sea, or through fish-farming and pond cultures. Java has at least 7.5 million artisanal fishermen operating in shallow coastal waters or raising fish and shrimp in small ponds. But as more people fish for a living, resource conflicts escalate and sometimes turn ugly. Before big shrimp trawlers were banned from Indonesia's coastal waters in 1980, gun battles raged between trawlermen and artisanal fishermen. In the 1970s, at least one hundred people were killed from such confrontations.

Before the trawlers were banned from near-shore waters, their fishing techniques earned some fleets the nickname "vacuum cleaners of the sea". Decades of bottom trawling in the Arafura Sea, in western Indonesia, has caused the seabed to look as if it has been crossed with motorways. The nets have scooped up everything in their path, turning the sea bottom into a desert.

Malaysians continue to fight each other over violations of fishing territory. "Because of the mismanagement of coastal resources, there has been a consistently downward trend in coastal fisheries in Malaysia since the 1960s," notes one top-level Malaysian fisheries expert, who wishes to remain anonymous. "Trawlers have often infringed in near-shore waters, coming into direct conflict with small-scale fishermen. A number of fishermen have been killed in such confrontations."

Fishermen in each state of Malaysia are allowed to fish only in their own waters, but there are many violators. Fish stocks are assessed regularly by the Fisheries Research Institute in Penang. Based on such assessments, licences to fish are allotted to each state. "The problem is, too many poor fishermen are operating without licences, and exploitation is getting out of control in some areas," explains the expert.

The other big problem for Malaysian fisheries is misplaced priorities. "We now have a big programme to exploit off-shore fisheries. But in my opinion, we should be putting more effort and money into managing coastal resources, instead of throwing it away on deep-sea fishing," he concludes.

In an attempt to come to terms with depleted fisheries and ruined coastal resources, the Malaysian government wants to reduce the number of coastal fishermen from 120,000 down to 70,000 over the course of the next decade. Those not allowed to fish any more, will be taught to raise fish in pond cultures. Since there are already many fish farmers in Malaysia, this solution may cause more problems than it solves. Moreover, the new ponds will most likely be built at the expense of the country's remaining mangroves.

One of the region's biggest problems is the widespread use of dynamite and poison to catch fish, especially on coral reefs. Virtually every country in South-east Asia has fishermen who insist on catching fish in this way. Sometimes entire villages specialize in these destructive fishing practices. Not only do they kill fish indiscriminately, but reef habitats are destroyed, depriving other marine life of shelter and food. And when the reefs are gone, so are the fish.

The Lingayen Gulf on the north-west coast of Luzon, the Philippines, is a classic example of over-exploitation, aggravated by habitat destruction from blast fishing and sedimentation from mining operations. Near the fishing village of Bolinao, "fishing areas are so depleted that we can't even assess how to optimize fisheries in this part of the Gulf, because we can't find any fish to study," explains Dr Liana McManus from the University of the Philippines in Manila.

Many reefs have been reduced to rubble. "In 1983 I didn't hear any underwater explosions while diving near Bolinao," observes Dr Helen Yap of the Marine Sciences Institute. "But in 1988 when I returned to do research, I couldn't count the number of explosions I heard during one dive. They were happening all the time."

Conflicts still occur in the Lingayen Gulf between trawlers and small-scale fishermen. Although trawling is not permitted within seven kilometres of the coast, many lay out their nets 2–3 km from shore. Years of trawling, and heavy fishing by the 15,000 artisanal fishermen in the Gulf, have left it impoverished of marine life. Whereas 20 years ago, fishermen could fish 25 days a month during the peak season, they now go out one week a month, and are lucky to catch anything. As resources run out, more of the poor fishing villages turn to illegal and destructive fishing methods.

San Roque (not the village's real name) is one of the poorest fishing villages along the Gulf. Of the 200 families in the village, 80 per cent of them earn their living directly from the sea. As fish stocks – and incomes – continue to decline, the entire village has taken up dynamite fishing. Although they realize that this is destroying the reefs and killing more fish than they can recover, they justify it by claiming that it is better to fish with dynamite than turn to other illegal activities like smuggling.

Often, in order to avoid arrest, the local police are bribed with fish or cash. In many cases, blast fishing takes place with the tacit approval of the police. "Since some of the local police have relatives in San Roque, they are reluctant to arrest anyone," explains one villager. "The heavy fines imposed on those caught could ruin the family."

San Roque is one of thousands of villages in Asia which, through poverty or greed, destroys the resource base it (and others) depends on. The immediate needs of today win out over the needs of tomorrow. Until alternative incomes can be provided, or the benefits of conservation accepted, the destruction will likely continue.

The Lingayen Gulf may be brought back from the brink of ecological disaster. There is now a coastal management plan, co-ordinated by the International Center for Living Aquatic Resources Management (ICLARM) in Manila. The Coastal Resources Management Project will set aside reserves, declare limited fishing areas, curb heavy-metal pollution from mining activities, and help with the establishment of fish and prawn farms in villages like San Roque. But if the plan is to work, it will require the active participation of all fishing villages along the Gulf.

There is one widespread social factor that is likely to mar attempts at the rational management of coastal resources: landlessness. Most rural households in Asia do not own the land they work. Around 78 per cent of Filipino and 85 per cent of Indonesian peasants own no land. When their soil wears out, or they are unable to pay rent, these poor families are forced off the land. Often, their only recourse is to head for the nearest coast.

Once these migrants arrive at the coast they usually add to the problems of the fishing communities already there. "Migrants are not restrained by webs of family and informal ties, nor do they have, as old fishermen sometimes do, a small plot of land to resort to when catches are poor," explains Dr Daniel Pauly, Director of ICLARM's Resource Assessment and Management Programme in Manila. "Hence, they will usually be among the first to use fishing techniques such as excessively-fine mesh nets, dynamite, cyanide and bleaches. In the Philippines and Indonesia these techniques now constitute the major fishing gear, especially in coralline areas."

Poverty is at the root of many of the problems afflicting South-east Asia's fisheries. "Solutions to fishing problems will be forthcoming only when the central issue, poverty itself, has been resolved," observes Dr Pauly.

Conservation

In some areas poverty has not prevented people from rescuing their resources. Conservation, even on a small scale, has worked wonders. On tiny Apo Island, off the southern coast of Negros (the Philippines), the local fishing community has managed to increase fish stocks by preserving just a small part of the coral reef that fringes the entire island. The sanctuary occupies a mere eight per cent of the 106-hectare reef, located a short distance off the windward side of the island. "Before the sanctuary was set up, it was difficult to catch fish, but now it is much easier," explains Jesus Delmo, President of the Apo Island Marine Management Committee. "Before the reserve came, we had to travel as far as Mindanao to fish, but not now. We can catch as much fish as we need around our own island. More people are fishing, but there are more fish." Even the bigger, tastier fish, like jacks, are beginning to return to the island's reef.

Before the reserve was established in 1986, the 90 fishing families on Apo were using dynamite and poison (sodium cyanide), as well as hooks and lines, to put food on the table. Apo islanders also employed the old *muro-ami* fishing technique, which involves scores of children free-diving down to the reef, pulling a large fixed net behind them. In order to frighten the reef fish out of their coral refuges and into the net, the coral is hit with big wooden mallets. The fish usually oblige by fleeing into the waiting net, but the coral is pulverized in the process. The waters around Apo Island were quickly becoming devoid of edible fish.

Since the villagers had no real concept of conservation, it took two social workers from Silliman University two years to convince them to allow the Marine Laboratory of the University to set up a pilot project on a small section of their reef. Between 1984 and 1986, the social workers spent two weeks of every month living on the island, quietly convincing the people of the value of conservation. Although the process was long and involved, it has paid off. The 200 people on Apo Island now see the reserve as their own and protect it from exploitation by fishermen from nearby islands, and scuba-divers. "We recently confiscated four spear-guns from some Japanese scuba-divers," says Jesus Delmo. "They were trying to spearfish in our reserve."

The "demonstration value" of the reserve has been enormous. "Fishermen on Apo Island have learned that by setting aside a tiny part of their reef for conservation, where no fishing of any kind is allowed, fish stocks on the rest of the reef have gradually increased," points out Louella Dolar, a faculty member of the Biology Department at Silliman University. The lesson has been passed on to, and accepted by, fishing communities on the neighbouring islands of Bohol, Siguijor and Cebu. So far, five other marine reserves have been established.

Other innovations have come from Professor Angel Alcala's group at Silliman University. Alcala was the first to develop artificial reefs made from discarded rubber tires and bamboo poles back in the late 1970s. They are now in use throughout South-east Asia by fishing communities which have lost their original coral reefs.

The Marine Laboratory at Silliman University is presently cultivating five species of giant clams in tanks. Heavily harvested throughout Asia, clams have disappeared from many reefs. The Marine Laboratory hopes to re-stock reefs with giant clams through controlled breeding and continuous monitoring during their critical first few years of life. Once the clams are seven months old they are transferred to cages placed on a section of protected reef at Apo Island. The local fishing village protects the clams because once they are three years old, islanders will be able to harvest half of them; the rest will remain as brood stock. If the project is successful, then the clams will be introduced to other Philippine reefs.

Dr Alcala is convinced that giant clams can be raised in economical quantities in tanks. If so, he may have laid the foundation for a new mariculture industry, one that will provide jobs and incomes in one of the most depressed areas in the Philippines.

Not far from Apo Island, on the east coast of Negros, another rescue operation is under way: the replacement of a denuded coastline with a young mangrove forest. This was begun by Wilson Vailoces, vice-president of the Tinaogan Fishermen's Association, but is now a full-scale community effort. "I started this activity on my own with no government assistance," says Wilson. "Now my neighbours are copying me and together we will reforest this entire coastline." Recently, the Government granted Wilson and members of his community a 25-year contract of stewardship over the land along the coast. "The protection of this resource is in our hands now," affirms Wilson. "And it is up to us to see that our children reap the benefits of our hard work."

The families who plant mangroves have the right to harvest some of the trees for their own use or to sell timber on the local market. Mangrove wood is highly-prized as building material, since it is very resilient and resistant to insect pests. But few trees are cut: they are regarded as a kind of savings account for the community and are most valuable where they are.

A hundred metres offshore, fishermen from the co-operative gather around a buoy in their out-riggers. They cast out their fishing lines – just simple hook-and-line fishing, no poles or nets. The fish are plentiful in this spot because Wilson Vailoces and his comrades have built an artificial reef out of bamboo poles to attract fish. "We needed to build a reef because our original reef was badly damaged from dynamite fishing and from the use of coral as building material," confesses Wilson. "Now we know better."

Unfortunately, many poor people on the island do not know better. Poverty, combined with large families, continues to force many subsistence farmers and fishermen alike to over-exploit their limited resources. The population growth rate on Negros stands at over three per cent a year, doubling their number every 23 years. And much of this growth is concentrated along the island's over-crowded coastline. In fact, the average fishing family in the Philippines has six or seven children, one child more than upland farm families.

The continued degradation of coastal resources is aggravated, and sometimes precipitated, by what happens inland. This part of the Philippines suffers from the same wanton destruction of upland forests that has plagued virtually every one of this country's 7,100 islands. Tropical forests, which covered nearly 75 per cent of the country in the 1950s, were reduced in area to less than 20 per cent by 1988. Over the past ten years alone deforestation has proceeded at the rate of about 105,000 hectares a year. Along the entire east coast of Negros only two peaks still retain a few hundred undisturbed hectares of tropical forest. The rest of the land is given over to coconut palms, pastures, agricultural fields and settlements.

The loss of forests has had an unintended side-effect: erosion of agricultural lands in the uplands. On Negros, erosion affects 50 per cent of the land area, a figure which is considered a national average. With less forest cover has come less rainfall. When it does rain, tonnes of soil and rock are gouged out of the hills and washed down to the lowlands and coastal areas. "We now have flash floods," complains Wilson. "And heavy rains wash soils down from the hills into our coasts. Sometimes huge boulders crash down the hillsides. In the end, nothing remains: no corals, no mangroves, only silt."

The fishing communities of Tinaogan continue to suffer from the mistakes of others. Learning to conserve and manage their own resources is only the first step. The next stage, much more difficult, is to change some local agricultural practices. "The farmers around here spray their rice crops six times a year with pesticides," explains Wilson. "Runoff from this poison goes into our shallow coastal waters and kills off many fish and shellfish." Particularly hard hit are crabs and shrimp fry. To make matters worse, spray guns are often cleaned out in streams which flow directly into the sea, bringing in more poisonous residues.

Wilson and other members of the Fishermen's Association have been trying to persuade local farmers to spray their crops less frequently and to use less harmful kinds of pesticides. So far, they have met with resistance. Farmers insist that the heavy doses of pesticides are necessary to insure higher yields by minimizing insect damage. Using safer pesticides, they claim, costs more money. Wilson is just as adamant. "The rice-farmers and fruit-growers here have to be educated to use poisons properly, not indiscriminately," he says. "The mentality of these farmers is spray, spray, spray. They have no concept of what their poisons do to the environment."

It has been a difficult struggle for Wilson and his neighbours. Not long ago, Tinaogan was very poor. The reef was destroyed, the mangroves cut down and fishing was becoming a precarious way of life, both for the established community and for the newly-arrived, displaced hill-farmers. Faced with disaster, the fishermen of Tinaogan began to fight to rebuild their resources.

A few of the peasant migrants have been assimilated into the fishing communities of Tinaogan. Others have taken employment as agricultural labourers, working for the wealthier farmers along the coast. But many have drifted into nearby towns, particularly Dumaguete City, where they squat in flimsy bamboo huts on the edge of town, taking any employment they can find. Some have turned to petty crime. For many of these "environmental refugees", life has become a daily struggle for basic necessities.

There is new hope at Tinaogan, however. Wilson Vailoces' mangroves are beginning to stabilize eroded shorelines, providing sanctuaries for a wide variety of edible fish and shellfish. Fish stocks are beginning to increase, and the people of the area are not so impoverished. By replanting their mangroves and building artificial reefs and fish-shelters, the small fishing communities of Tinaogan have learned to manage their resources. The battle is far from over, but a good beginning has been made.

"The education we have received, we will pass on to our children and grandchildren," says Vailoces, "so the investment is worth it. We have learned the hard way how important it is to keep our resources intact." As fish shoal in the shallows, near his mangroves, Wilson smiles: "We depend for our very lives on the sea, so we have to take care of our fisheries and our resources. It's all we have."

Coastal subsidence

Besides pollution, Manila and Bangkok share another unsavoury problem. Both cities are sinking. Excessive withdrawal of groundwater for municipal and industrial use has caused surface land to sink, a process known as subsidence. Saltwater is now intruding into underground aquifers, making them unfit for municipal or industrial use.

Manila Bay has been rising at the constant rate of two centimetres a year for the past 15 years. Scientists are not certain if this is due to subsidence or the fact that huge portions of the Bay have been filled in for land reclamation projects. Whatever the cause, it has city managers worried.

North of Manila, along the coast, fish farms and rice fields are suffering from salt water intrusion into their shallow groundwater supplies. "The entire coastal environment of this region could be permanently altered to a more salty one," explains Ricardo Biña of ICLARM in Manila. "We don't know the consequences of this process. It could be catastrophic."

Bangkok has begun to drill much deeper wells in an effort to supply its citizens with untainted fresh water. But drawing water from deeper wells will not stop the city from sinking. Another reason for the deeper wells is that Bangkok's water supplies are contaminated with raw sewage which seeps into aquifers from unlined sewage canals. It's a problem that is regarded as far more serious than surface subsidence.

Grappling with these immediate problems has not allowed city planners to contemplate another emerging crisis: the effects on their economies of rising sea-levels from global climate change. Global warning means that the oceans are expanding (water expands when heated), possibly to the extent that low-lying coastal areas will be inundated. What this means for Asia and the Pacific is frightening to imagine. A sea-level rise of as little as half a metre could displace millions of people, most of them poor. Where would they go in an already over populated area?

Singapore's success story

Singapore's waterfront used to be an unsightly scene. Floating scum and rotting garbage covered the harbour and collected around boats. Pig and chicken farms discharged their wastes directly into the Singapore River. Boats dumped their wastes into the harbour. Food vendors washed out their stalls on the streets; the slops washed into the Kallang Basin. All of this pollution eventually collected in the marina and harbour, where it festered.

In 1977 the government decided to rid the city of this eyesore and public health menace. The catchments of the Singapore River and the Kallang Basin would be cleaned up. Ten years and $100 million later, the entire waterfront has been restored.

The job was complicated and time-consuming. Sewage lines had to be built to accommodate some 12,000 unserviced homes and industries. A new two-stage sewage treatment plant was built. Hundreds of pig and chicken farms had to be moved. Around 5,000 street vendors were relocated and given new stalls with proper drains connected to the sewer system. The harbour was dredged.

"Once we removed the main sources of pollution, the rivers cleaned themselves," points out Ms Kuan Kwee Jee, Deputy Director of the Ministry of the Environment. Still, it takes not only money but strong political motivation to carry through with such a massive programme. Perhaps only Singapore, with its authoritarian government, could have achieved such results. Nevertheless, Singapore's success story stands as an example of what can be accomplished.

The East Asian Seas Action Plan

Asian governments, committed to developing their economies, traditionally view environmental protection as a luxury they cannot afford. Although most governments of the region have established bureaus or ministries of the environment, they usually do not give them much authority. Most environment offices are compelled to operate on tiny budgets, with too few experts. Programmes are swallowed by red tape; co-ordination with other key ministries is lacking.

But attitudes towards environmental and resource management are changing. When UNEP suggested forming a regional seas Action Plan, the governments of the region agreed. South-east Asian countries saw a regional organization as a way to deal with common resource problems.

After a series of preliminary meetings, the East Asian Seas Action Plan was adopted in April 1981 at an intergovernmental meeting in Manila. The Action Plan aims to provide a framework for a comprehensive and environmentally-sound approach to coastal-area development. In order to oversee its implementation the Co-ordinating Body on the Seas of East Asia (COBSEA) was established. This co-ordinating body, with representatives from each participating government, has overall authority to identify priorities and determine the budget. At the request of the participating governments, UNEP is providing Secretariat services for the Action Plan, including management of the Trust Fund.

The fund was set up to cover the costs of the programme. So far, however, contributions to it have only given the Action Plan an annual budget of about $100,000, though UNEP's contributions have totalled $2.2 million. UN officials see the lack of financial commitment on the part of the region's governments as a problem that must be resolved if the programme is to achieve its objectives.

At the sixth meeting of COBSEA in 1987, government representatives proposed a long-term strategy for managing the region's marine and coastal resources. At the same time, UNEP helped launch the Association of South-east Asian Marine Scientists (ASEAMS), in order to involve the science community and provide a mechanism for inter-regional co-operation. Since its inception in 1987, ASEAMS's membership has grown to around 120 and a newsletter is published and widely distributed. The Association held its first international symposium in February 1989, with another scheduled for 1991. Recently, ASEAMS prepared a UNEP-funded report on the likely effects of sea-level rise on the coasts and resources of South-east Asia.

The big issues may not have been tackled yet, but the Action Plan's scientific base has been considerably broadened. The programme has trained 500 technicians and scientists from East Asia. Data collection has improved, along with data exchange. The five countries of COBSEA are beginning to collaborate on a number of projects, a difficult task before the advent of the Action Plan. "Before the initiation of the programme, the countries of the region had no integrated activities to address the problems of the marine environment," states Ed Gomez. "Through this action plan, scientists and environmentalists concerned with the protection of the marine environment have been able to broaden their focus from a national to a regional perspective."

9. South Asia

South Asia is one of the poorest regions on earth. The average annual *per capita* income for India, Pakistan, Bangladesh, Sri Lanka, and the Maldives is no more than $300. At least a quarter of all Indians live along the coast; a third within 100 kilometres of it. The poorest people of Bangladesh live in over-crowded slums built on stilts in the delta of the Ganges and Brahmaputra Rivers. Here they are at the mercy of flood waters from the land and storm surges from the sea.

Pressures on South Asian coastal resources are tremendous. The effects of these pressures are familiar. Mangrove forest, which stabilize shorelines and deflect storms, are being exploited on a large scale for firewood, tannins and timber, and are also being converted into brackish-water fish and shrimp ponds. Coral reefs face assaults from sedimentation due to coastal construction, soil erosion and dredging of harbours. Coastal forests have been replaced with agricultural land and towns. Over-fishing is a chronic problem along the entire coast from Pakistan to Bangladesh. Over eighty per cent of all wastes are flushed into coastal waters untreated. Many beaches are used as garbage dumps, while solid wastes are often disposed of in shallow coastal waters.

Space is at a premium along the coasts of Pakistan, India and Bangladesh. Coastal populations are increasing, often at more rapid rates than inland. With growing numbers of people comes the proliferation of slums and squatter settlements. Over half of Bombay's population of 10 million live in slums, as do nearly 40 per cent of Calcutta's.

The region's coasts are booming with activity. Unfortunately, nearly all of it is unregulated and unplanned. "The mentality prevailing along the south coast of India is something akin to what the wild west in America must have been like in the last century," says one Indian scientist. "Anything goes, and the guy with the biggest stick is always right."

Land and sea: an uneasy alliance

The north-central part of the Indian Ocean is divided into four distinct parts: the Arabian Sea, the Bay of Bengal, and the Andaman and Laccadive Seas.

The continental shelves vary in width from a few hundred metres to several hundred kilometres, allowing for a diverse fishing industry. Upwellings of nutrient rich water in the Arabian Sea make it a particularly valuable fishing ground.

Perhaps in no other region of the world is the give and take of land and sea so visible. The region is racked by geographical and meteorological extremes. Seasonal monsoons flood coastal areas, swamping lowlands and river deltas. On the other hand, so much sediment runs off the land every year – about 1.6 billion tonnes – that new islands are regularly formed in the huge deltas of the Ganges and Brahmaputra Rivers (in the Bay of Bengal) and at the mouth of the Indus River (in the Arabian Sea). "These two marine areas occupy three per cent of the world oceans, but receive nine per cent of the global runoff," writes Indian scientist Dr Sen Gupta.

Agriculture remains the mainstay of the region's economies, but fishing, tourism, mining and industrial development are growing.

All of the region's countries have declared 200-mile economic exclusion zones. Sri Lanka's EEZ, comprising 1.2 million sq km, is twenty times its land area. With fishing limited to near-shore waters, Sri Lanka is hoping to sell its deep water fishing rights to India and Pakistan, whose tuna fleets range across the Indian Ocean.

Crowded coasts

People remain the region's greatest resource as well as its greatest problem. On the island of Male in the Maldives, 40,000 live on one sq km of land. The city has no public sewer system. Instead, the people of Male use the Gifili system, digging a small new hole for every use. The island's freshwater lens, situated only a few meters deep in the sandy soil, sits precariously atop a wedge of saltwater. Not only are the island's limited supplies of groundwater contaminated with faecal matter, which seeps into it from all the latrines and shallow holes, but over-use is causing saltwater to intrude into the freshwater lens. One result is that diarrhoeal disease and intestinal disorders are epidemic.

With nearly a billion people living on a little more than 5 million sq km, South Asia's population density is high; it averages almost 200 people per sq km. Along its coastline, population densities approach 400–500 per sq km.

The state of Kerala, wedged into the south–west tip of India, has one of the highest population densities in the country. So many people live in this state that it has been described as "one continuous village". But the density of the fishing population along the coast is twice that of the rest of Kerala, at around 1,000 people per sq km.

Urban areas are expanding throughout South Asia, as is the case in virtually every other part of the world. The region's largest cities are all located on or near coasts: Karachi, Hyderabad, Bombay, Ahmadabad, Mangalore, Madras, Vishakhapatnam, Calcutta, Dhaka, Chittagong and Colombo.

Although overall population growth rates are slowing down – averaging a little over two per cent a year – the growth of cities and towns continues to escalate. At current rates of growth, cities like Bombay will double their populations every 17 years. This is a sobering thought, considering that Bombay's slums and squatter settlements already contain nearly five million people. If trends continue, planners predict that 75 per cent of the city will be forced into slums by the turn of the century.

Naven Panjwani, in *The State of India's Environment 1984–85*, describes the shocking conditions in Indian slums:

> A survey of about 4,000 households in nine slums in Bombay reveals that nearly 40 per cent have 2–4 persons packed into one room, another 35 per cent of households have 5–9 persons crammed into one living room, and one per cent have 10 or more people in one room. No house has a private toilet. A quarter of the households do not even have access to community toilets and use the open spaces around the slum for defecation. Over a third have no drainage facilities and another 40 per cent have uncovered drains.

At last count, over 3 million people were living in 3,000 slums around Calcutta. Household slops and wastes are dumped into open latrines and drainage ditches. Respiratory ailments and gastro-intestinal diseases afflict almost three-quarters of all slum-dwellers. Most sickness is blamed on the acute lack of sanitary facilities and access to clean water. Every morning, one of the most common sights across

South Asia is scores of people squatting over drainage canals or by stagnant streams, while a few metres away others bathe themselves and wash their clothes.

As urban populations continue to grow, city authorities find it impossible to provide adequate services. The Calcutta Metropolitan Development Authority has tried since the early 1970s to provide a sanitary latrine for every 25 slum dwellers, a potable water tap for every 100 people and proper drains. Even where this modest goal has been achieved, the facilities have not been maintained. Drains are clogged with rotting garbage, latrines overflow with excrement, and taps run dry. Funds to keep the services working often go into the pockets of unscrupulous contractors.

The World Health Organization estimates that only half of the people of South Asia, around 500 million, have access to clean water and adequate sanitation. Even where sewage systems exist, their contents are pumped untreated into coastal waters or into local rivers and streams. Eventually, most of the wastes of one billion people find their way to the region's coastal areas.

Coastal pollution

Sewage treatment in South Asia is almost non-existent. All of Bangladesh's sewage, for example, is flushed directly into the sea or goes there via the Ganges and Brahmaputra Rivers. Only a tiny fraction of municipal and industrial wastes from Pakistan and India receive any form of treatment before being dumped in coastal waters or into rivers and canals. Bombay discharges around 365 million tonnes of untreated sewage and municipal wastes into the sea every year. Similarly, Calcutta dumps close to 400 million tonnes of raw sewage into the Hooghly Estuary. Karachi, in Pakistan, pumps roughly 175 million tonnes of sewage and industrial wastes into the Arabian Sea annually.

High bacteria counts from untreated wastes are found in nearly all near-shore waters along South Asia's urbanized coastline. Shellfish beds near Madras are contaminated with human wastes, making them unsuitable for consumption. Tons of the mussels are harvested anyway and sold to the city's poor. Sri Lanka's shellfish are contaminated, especially near the capital Colombo. The Manora Channel in Karachi Harbour reeks from hydrogen sulfide fumes, its black, bubbling waters filled with industrial poisons and untreated sewage. Dead fish are a common sight in nearby waters. Although

Karachi has two single-stage sewage treatment plants, their capacity is constantly overwhelmed.

The World Bank is providing Bombay with loans to assist in building a sewage treatment network for the city. Plans call for the outfalls to be stretched three km offshore, but it may be a decade before the system is functional.

Coastal seas are also garbage bins. Some 34 million tonnes of garbage and solid wastes are dumped into India's coastal waters every year, ruining seagrass beds and coral reefs. In Sri Lanka, the ocean is the main receptacle for solid wastes.

Pollution of near-shore waters from chemical pesticides and fertilizers is a serious problem in parts of India, Pakistan, Bangladesh and Sri Lanka. Agricultural communities along India's coast spread 5 million tonnes of fertilizers a year on their fields and use 55,000 tonnes of pesticides. Residues from these chemicals end up in coastal ecosystems, poisoning marine life and adding to the problems of eutrophication (lack of oxygen) initiated by millions of tons of untreated sewage. An Indian marine scientist remarks that "some of our coastal areas are eco-disasters". Pakistan's Indus River is packed with pesticide residues. Sri Lanka uses so much fertilizer on its fields that groundwater reserves are becoming fouled with nitrates. In some areas water-wells are so full of nitrates and other agricultural chemicals that runoff from them kills nearby vegetation.

Industrial pollution is limited to major urban centres. But untreated industrial wastes pose an increasing threat to coastal areas. In the Bangladeshi city of Chittagong, a petroleum refinery's wastes contaminated the city's water supply, making it unusable. Water is brought in by tanker and pumped from neighbouring areas. On Sri Lanka, industrial effluents from food-processing industries and slaughter houses have killed fish in the Kelani River and turned the Valachchenai Lagoon dark brown. The waters off Bombay are dangerously polluted from heavy metals. Cadmium levels in the water column are among the highest ever recorded (80 microgrammes per litre). There are widespread fears that as the region develops its industrial infrastructure, industrial pollution will accelerate, compounding the problems posed by raw municipal wastes.

The sub-continent's rivers, especially the Ganges, Brahmaputra and Indus bring in millions of tonnes of sewage and industrial wastes. The Ganges in particular is full of toxics, including decomposing bodies tossed into it along most of its length: chronic fuelwood shortages that afflict most of India mean that wood to burn bodies is expensive and hard to get. The poor simply cannot afford it. At just one Indian city, Varanasi, about 10,000 half-burned bodies are pushed into the river each year, along with 60,000 carcasses of cows, dogs and buffaloes. Downstream, the river's waters contain a frightening assortment of toxins and disease organisms.

In a bid to clean up the Ganges, the Indian government has allocated $195 million over the next five years. Sewage treatment plants will be built, along with industrial-waste treatment facilities. The carcasses of animals will not be allowed to be dumped in the river at all and proper cremation of bodies will be mandatory. In urban areas large cattle sheds will be built and the dung collected for use in local biogas plants and as fertilizer. Comprehensive as the clean-up programme is, it will take some years before any change is noted in the river's water quality.

Oil pollution

South Asia lives on the same oil highway that passes through South-east Asian waters. In 1985, around 232 million tonnes of Middle East crude were transported along the main tanker route, which stretches across the Arabian Sea, through the Maldives and into East Asian waters.

Oil spills occur, but the real problem, as elsewhere, is the routine discharge of dirty ballast waters, bilge slops and tank washings from oil tankers and other ships plying the shipping lanes in the Indian Ocean. It has been estimated that roughly five million tonnes of oil is slopped into the Arabian Sea every year, while the Bay of Bengal gets 400,000 tonnes.

As a consequence of this practice, tar-balls and oily residues are a common sight on most beaches bordering the Indian Ocean. Some of south India's beaches are permanently littered with tar. Although some 5,000 tankers pass within five km of Sri Lanka's coast every year, so far no serious accidents have occurred – but time is not on Sri Lanka's side. Coral reefs have been covered with oil residues in the Andaman and Nicobar island archipelagos in the Andaman Sea, reducing income from tourism. So far, there is no region-wide mechanism in place to deal with oil spills and the routine discharge of oily wastes by ships sailing through the Indian Ocean.

Coastal sedimentation

Most of the region's rivers run brown sediment from agricultural lands and denuded watersheds. Much of this sediment ends up being dumped into coastal waters. The Brahmaputra River alone carries roughly 500 million tonnes of sediment a year into the Bay of Bengal. Bound to these sediments are heavy metals and chemical residues from industrial, mining and agricultural activities.

Coastal waters are often muddied by all this debris from the land, reducing fish stocks by soiling breeding habitats. Upland soil erosion in Sri Lanka has resulted in the siltation of important lagoonal areas. In some cases, silt has built up across the mouths of lagoons, cutting them off from the sea and destroying breeding and nursery areas for fish and shellfish.

In those areas where mangroves stabilize some of the sediment and filter out pollutants, the problems caused by sediment overloading are greatly minimized. But aside from the Sundarbans (at the mouth of the Ganges and Brahmaputra), the largest tract of mangroves in South Asia, much of the region's coasts have been denuded of their protective mangrove barriers. And once the mangroves are gone, coasts are ravaged by storms and coastal land is lost to the sea.

Siltation of coastal areas does have a positive aspect: it builds up land. The extensive mangroves of the Sundarbans are a product of this natural process. Normally, mangrove forests would spread seaward, colonizing many of the new islands currently being built up by erosion sediment dumped into the Bay of Bengal by the Brahmaputra and Ganges Rivers. Between 1972 and 1977, 3,600 hectares of land were added to the Bay. Unfortunately, pressure for coastal land has reached crisis proportions in Bangladesh and India. Tens of thousands of poor landless fishermen and peasants are forced onto any new spits of sand, known as *chars*, that appear above the sea. Mangroves do not have a chance to extend their reach and stabilize the land. The absence of mangroves on these newly-formed islands translates directly into human tragedy when tens of thousands of people are swept away by tropical storms. In some cases even the islands themselves disappear.

Mangroves and coral reefs at risk

Healthy mangroves and coral reefs are essential components of coastal ecosystems in the Indian Ocean. If managed properly they provide employment as well as food, fibre, fodder and building materials.

The Sundarbans, that vast mangrove swamp sprawling between the deltas of the Hooghly and Ganges/Brahmaputra Rivers, covers around 6,000 sq km. It is one of the largest mangrove areas in the world, and has been worked for decades by coastal peoples in both India and Bangladesh. Currently, this vast forest supports around 300,000 people who make their living by selectively harvesting mangrove timber and raising fish and shrimp in pond cultures. In recent years, however, excessive harvesting of mangrove wood and conversion to fish ponds has degraded parts of the Sundarbans. The area is now more at risk from storm surges and tidal floods.

India may have upwards of 700,000 hectares of mangrove swamps along its 7,000 km of coastline. But their status remains uncertain, as no real inventories have been conducted. And thousands of hectares have already been converted into brackish-water shrimp and fish ponds, agricultural land, industrial estates and housing units.

Extensive deforestation of mangroves in the Gulf of Kutch has resulted in the sedimentation of nearby coral reefs. Mangroves have also been cleared around Bombay and the Cochin Backwaters.

Pakistan's mangroves occupy a fraction of their former range. The only undisturbed stand is at the mouth of the Indus River. Sri Lanka has perhaps 6,000 hectares of mangroves left, but they are being exploited on a large scale for fuelwood and fish ponds.

In addition to their value as fish-breeding centres, coral reefs perform other important functions. Like mangroves, they reduce wave action and help prevent shoreline and beach erosion. If coral communities are destroyed, the consequences can be as bad for the land as they are for fisheries. On the south-west coast of Sri Lanka, near the town of Hikkaduwa, coral mining for construction purposes (for lime, calcium carbide and cement) has removed some 75,000 tonnes of coral a year. As a result, the "beach" has moved 300 metres inland, whittled away by direct wave action over a 50-year period.

The best reefs are found around the Maldive Islands, hundreds of which are uninhabited. Remoteness has saved many of them. The coral reefs near inhabited islands, particularly around the main island of Male, have been destroyed by coral collectors and mined for the building industry.

The one reef off Bangladesh has been blasted apart by dynamite fishermen. The coral islands and atolls of the Andaman and Nicobar archipelagos are relatively free from exploitation, except for coral collection for the tourist trade. However, they lie close to shipping lanes and some of the fringing reefs are routinely covered with oil pollution. These reefs have also been invaded by the destructive crown-of-thorns starfish.

Because the Indian Ocean has areas of nutrient-rich upwellings where fish congregate, coral-reef fisheries are very localized, contributing little to the commercial catch.

Fisheries

The coastal waters of the Arabian Sea and the Bay of Bengal are rich in fish and shellfish, thanks to upwellings and favourable currents. The countries of the region collectively gather 2.4 million tonnes of fish and shellfish a year from this part of the Indian Ocean.

India and Pakistan both have deep water fishing fleets. But thousands of tonnes of fish are taken by small-scale commercial and artisanal fishermen in each country. In many areas, particularly off India's south coast, conflicts have erupted between subsistence fishermen and trawlermen. Several hundred have been injured in these confrontations. As near-shore fish catches continue to decline – due in large measure to over-fishing and pollution – such conflicts will likely continue. Trawlers also continue to fish within three miles of the shore, angering subsistence fishermen, who lack motors and bigger boats which would allow them to go further off-shore to fish.

India has around 20,000 mechanized trawlers, in addition to 75 deep sea fishing vessels. Another 200 trawlers will be added to the fleet over the course of the next few years. But little is being done to help small-scale, artisanal fishermen upgrade their equipment and skills. For the 10 million or so subsistence fishermen along India's coastline, making a living from the sea is becoming more difficult every year.

Yields from cultivated fish and shellfish continue to expand. India harvests around 40,000 tonnes of pond-raised fish and shrimp a year; Sri Lanka close to 100,000 tonnes. More mangroves in the Sundarbans are being cleared for fish ponds, a development that could bring decreased off-shore fish catches, particularly prawns and shrimp, which are dependent on mangroves for critical stages in their life cycles.

Few studies of sustainable yields in the north-central Indian Ocean have been carried out, so it is difficult to tell if over-fishing of commercial stocks is a problem. Certainly many near-shore fisheries are chronically over-harvested. The governments of Pakistan, India and Bangladesh will have to come to terms with the needs of small-scale commercial fishermen if they want to implement realistic coastal-management plans.

Marine protected areas

The region has done little to conserve marine resources. Only India has an established system of marine parks and protected areas that qualify as something more than "paper parks". So far, 21 different sites have been set aside, most of them on the relatively unspoiled islands of the Andaman and Nicobar archipelagos.

Pakistan has granted some protection to the mangrove swamps of the Indus Delta. Sri Lanka has had little success in establishing marine reserves. The Hikkaduwa Marine Sanctuary was established in 1979, but has received no enforcement or management. Meanwhile, coral mining has eroded shorelines in the immediate vicinity of the sanctuary. The Maldives, which draw most of their income from fishing and tourism, have yet to establish any conservation areas. Since many of their coral islands are uninhabited and remote, a number of them could be made into sanctuaries and reserves. But as yet no studies have been carried out to assess their potential as protected areas.

Waiting for an action plan

An action plan for this region has been drafted and is now under consideration by South Asian governments. UNEP officials hope that it will be formally adopted in 1991, along with a convention and associated protocols combating oil spills.

"In addition to UNEP's catalytic role in getting these countries together, we have also initiated a number of preliminary programmes," points out Dr Reza Amini, former Programme Officer for UNEP's Oceans and Coastal Areas Programme Activity Centre in Nairobi. "IUCN will be investigating suitable sites for marine parks, and there is a public awareness campaign in the works as well." So far UNEP has invested half a million dollars into the region in an effort to get the action plan started.

It is obvious that these five countries desperately need a plan. Legislation controlling pollution from towns and cities and curbing harmful industrial effluents is badly needed. But even more important, coastal land-use strategies have to be worked out with the co-operation of local communities. Without effective management of coastal areas, resource conflicts will escalate and there will be fewer resources to sustain increasing numbers of people, millions of whom are poor.

10. Eastern Africa

The countries of Eastern Africa include the mainland nations of Somalia, Kenya, Tanzania, and Mozambique, and the islands of Madagascar (the Malagasy Republic), Mauritius, the Comoros, Réunion, and the Seychelles. With the exception of Réunion, which is part of France, all of these nations have foreign debts. Collectively, these countries owe $14 billion to foreign banks and development agencies. Although this sum is not nearly as much as Asian countries owe, servicing these debts is a severe drain on their economies. Tanzania's long-term public debt constitutes a full 85 per cent of its GNP; for Madagascar the figure is 106 per cent; for Somalia 75 per cent; and for Kenya 52 per cent.

One result of debt is shifting government priorities and unplanned development. Coastal development in all of these nations is accelerating, particularly in Kenya and Tanzania, as people move to the coast from the impoverished uplands. Kenya's coast from Diani north to Malindi is also being developed for tourism, displacing fishing villages.

As coastal crowding continues, land is degraded and mis-used. Coral reefs are damaged from blast fishing and sediment run-off from the land. Deforestation along the coasts, including mangrove forests, is a problem in some areas, as fuelwood becomes scarce and harder to find. Pollution around cities and industrial zones is on the increase. Behind the postcard facade of pristine lagoons, sheltered coves, and dazzling white-sand beaches, East Africa's coasts are threatened.

The region's colonial past is partly to blame. During long years of colonial exploitation, little was done to protect the environment or conserve resources. Somalia was occupied by the Italians, then the British, finally gaining independence in 1960. Kenya was a British colony until 1963. Tanzania was taken over by the Germans under Bismarck in 1898 in order to secure trade concessions, but after World War I the British took over until 1961. Mozambique was

Portugese until 1975. All of the island states were initially colonized by the French, but most of them were also British at some time during the last century. The French annexed Madagascar in 1896, with the connivance of the British, remaining in control until the country became independent in 1960. The Comoros became independent from France only in 1975. The English ended up with the Seychelles, finally allowing the creation of an independent state in 1976.

Most of the mainland countries have had independence now for more than 30 years. But the economic crisis means that resource and conservation issues are placed after development on the political agenda. Not one East African country has implemented comprehensive land-use policies. "And real planning capability simply doesn't exist," complains one Kenyan scientist. "In our rush to develop, we have forgotten resource management completely."

The long history of domination by foreign powers has left the countries of East Africa with some bitter legacies, but it has also provided them with a common ground. Realizing that a regional programme was necessary to come to terms with the growing problems of their coastal areas, they joined together in 1985 to endorse the Eastern African Action Plan.

A geography of poverty

As the travel guides state, Eastern Africa is a region of contrasts. Living amidst some of the most spectacular geography found anywhere are some of the world's poorest people. Stunning scenery abounds, from Kenya's Rift Valley to Mount Kilimanjaro in Tanzania to the lush flood plains of Mozambique. But it is the struggling poor of the region who will determine how that geography will look in ten years' time and beyond.

The narrow coastal plains contain some of East Africa's most fertile land and receive the most rainfall. The land here is green and richly productive. But much of the best land along the coast is worked by large-scale agricultural plantations growing only one or two crops, like sisal, sugar cane, pineapples, coffee, cotton, coconuts, or bananas. Smaller family-owned farms, which traditionally grow cassava, maize, cashews, coconuts and spices, are being pushed aside.

The cruel irony is that despite ample land for farming and grazing, East Africans have trouble feeding themselves. Much of what is grown is exported. And for many reasons – mostly political – the prices paid to peasant farmers are kept artificially low. The "pauperization of peasant farmers" has driven many of them from the land. Often men have migrated to cities and towns to find wage employment, leaving women and children to tend the family farm. "Not only have we squandered natural resources but we are squandering our human resources as well," notes one Tanzanian scientist.

Throughout the entire region, the coastal plain is only 10–20 km wide. This means that people and pollution are concentrated in narrow coastal strips. Except for arid Somalia, the coastline has many towns and fishing villages, industries, farms and tourist resorts.

Shallow continental shelves extend for only a few kilometres offshore before plunging down to 2,000 metres. These deep basins eventually taper off to 4,000 metres, the same average depth as the Pacific Ocean. Upwellings are seasonal with the monsoons. The rainy season, which lasts from April to August, is the prime fishing time, since the Somali Current flows counter-clockwise back towards the coast. With it come tropical cyclones, but also large quantities of nutrients and fish.

The island states of Madagascar and the Seychelles are remnants of the ancient super-continent known as Gondwanaland. The other islands – Comoros, Mauritius and Réunion – are all products of volcanic activity and are much younger.

Most of the islands are surrounded by fringing reefs, which create beautiful, shallow lagoons, ideal for tourism. Coral reefs are also land-builders. Without them, the Seychelles would lack 47 per cent of their total area, as well as about half of their 48,334 sq km of submarine banks. But the region's reefs are suffering from the pressures of exploitation for cement production, coral collection and fishing. And tourist developments more often than not ruin the resource which originally makes them attractive.

The fastest growing populations in the world

The sustainable management of resources is not helped by rapid population growth rates. Officially the population of the region is no more than around 60 million (as of 1985). But it could very well be double that figure – in many countries a proper census hasn't been carried out in years. All of the region's mainland populations (and Madagascar) are growing at nearly three per cent a year; Kenya's birth rate remains among the highest in the world, doubling the population of the country every 15 years.

Although the majority of Somalis, Kenyans and Tanzanians live in the uplands, a full 75 per cent of Mozambique's population lives along its 2,400 km of coastline. Similarly, most Malagasys live along the coast, as do the populations of the small island states.

The weight of agricultural and grazing pressures on fragile uplands throughout the region has contributed to land degradation and widespread soil erosion. As populations grow, good agricultural land shrinks. The best land in Kenya, for example, has been sub-divided again and again. The amount of land available per person has fallen from 0.40 hectares in 1969 to 0.20 hectares today. Poor farmers are forced to use marginal land, which swiftly deteriorates, or else move to cities and towns. Kenya's capital, Nairobi, grew 600 per cent between 1950 and 1979, and most of that growth was concentrated in slums and squatter settlements around the fringes of the city. Not surprisingly, one of Kenya's immediate development priorities is slowing population growth.

With the uplands incapable of supporting growing numbers of people, many migrate to urban areas along the coast. Since there are few other opportunities for employment, they turn to the sea in an effort to make a living.

This trend is painfully visible along Kenya's coast, both north and south of Mombasa. As more people move to the coast, they are forced to squat on whatever land they can find. This often brings them into direct conflict with land developers and the government.

During April 1989, 500 people squatting near the fishing village of Shariani, in Kilifi District north of Mombasa, were evicted from land they had been living on for over 30 years. Their houses were pulverized by bulldozers. In another incident the same month, squatters, angry at being forcibly evicted from their land, attacked police in the village of Diani, south of Mombasa. At least nine people were seriously injured in the ensuing riots. Kenya's coastal lands are in a state of crisis as squatters and the government continue to clash over land rights. Similar problems confront Tanzania as the country struggles with problems of coastal development.

The staggering growth of coastal cities and towns is one result of the failure of agrarian policies and the inability of East African governments to come to terms with the crisis of their agricultural lands. Dar es Salaam (Tanzania), Maputo (Mozambique), and Mogadishu (Somalia) have all witnessed extremely rapid and unplanned urban growth. Most of the new arrivals, forced off their land, end up in dismal disease-ridden slums or squatter settlements lacking clean water and sanitation.

Unfortunately, government planners, anxious only for development, have not yet made the connection between degraded uplands and over-populated, degraded coasts.

In some countries, such as Tanzania, government resettlement schemes have back-fired. The creation of *ujamaa* ("working together") villages has created chaos, rather than stability. In many cases these villages were not sited properly to begin with, so their new inhabitants could not make a living from the poor soils. Necessities like potable water and sanitation facilities were often overlooked. Dissatisfied with the arrangements in these *ujamaa*, many peasants simply migrated to coastal cities like the capital Dar es Salaam. Huge urban squatter settlements soon sprang up around the fringes of the city, prompting a crisis in municipal services. When authorities forcibly evicted thousands of squatters in the early 1980s, and sent them back to the interior, more displaced peasants took their place.

Another relocation programme forced people out of the traditional rice-growing region of the Rufiji Flood Plain and into the surrounding hills. The peasants, not used to growing anything in the drier uplands, could barely produce enough food for their own use. Meanwhile, rice production on the flood plain collapsed. Many of the peasants forcibly moved into the uplands eventually abandoned their small parcels of land and migrated to urban slums and squatter settlements in Tanzania's coastal cities and towns.

Rural resettlement schemes in Mozambique have produced similar results. The concentration of rural people into village centres has denuded lands around the villages, creating deserts where once there had been productive soil.

Well-meaning, but wrong-headed, agricultural policies have contributed to the impoverishment of soils and upland watersheds, and hence to Africa's food crisis. Improper grazing practices in the northern areas of Somalia, Kenya and Tanzania have caused the desert to advance.

The debt crisis is cited as a serious problem, but many knowledgeable Africans think it should not be used as an excuse for failing to manage resources. "We shouldn't confuse debt repayment with environmental deterioration," points out Dr Walter Lusigi, programme specialist for UNESCO's regional office in Nairobi. "We can survive debt, we cannot survive the loss of our environment. Rich or poor alike need to preserve their environment, or we all end up with nothing."

Resource degradation

As their resource base deteriorates, too many people throughout East Africa are ending up with nothing. One of the main threats to coastal resources remains the widespread degradation of uplands. Deforestation, poor farming practices and over-grazing have resulted in the loss of soil cover over much of the region. Barren soil, exposed to wind and torrential rains, quickly erodes. What doesn't erode is baked hard by the sun. The major rivers of East Africa and Madagascar run red with sediment washed out of the watershed. Most of this sediment ends up in shallow coastal waters, where it clogs estuaries, and smothers mangrove forests, seagrasses and coral reefs.

The Galana-Sabaki River in Kenya brings sediment washed out of the uplands into coastal waters near a marine park at Malindi. As a result, during the rainy season the waters around this resort town are choked with sediment. Near-shore waters turn a dark brown, repelling tourists and reducing local income. The beach at Malindi, once a major tourist attraction for Kenya, has expanded seaward by some 500 metres due to sediment deposits.

So much sediment has been dumped at the mouth of the Zambezi River in Mozambique that the delta is shrinking under the weight. The resulting rise in sea-level has caused extensive coastal erosion and salt water is intruding 80 km up the river, altering the entire ecology of the delta.

Madagascar is eroding away. Its denuded uplands are victims of excessive fuelwood-cutting, slash-and-burn agriculture, and over-grazing by sheep and goats. As one flies over the central part of Madagascar, the most notable feature is its barren hills, stripped of vegetation, punctuated by huge craters gouged out by water erosion. Nearly every river on the island is clogged with sediment, turning them a red-orange colour. So much sediment was dumped into the coastal waters of Mahajanga, that the city's port became completely inoperable. A new one had to be built further down the coast.

In some upland areas of Madagascar, as much as 250 tonnes of soil per hectare are lost every year. Soils there tend to be erosion-prone, a condition aggravated by tropical cyclones which can deluge the island with up to 15 millimetres of rain in 15 minutes. Expressed in monetary terms, the average loss from erosion reaches about $100–150 per hectare per year, the equivalent of 70–100 per cent of the average Malagasy's annual income. The government has counter-attacked with soil conservation programmes, resettlement schemes, and tough legislation, but the problems are far from solved.

The silt-choked Shebelle River transports thousands of tons of Ethiopian sediment to the Somali coast where it ruins beaches and clogs estuaries. Entire marine communities of plants and animals have been wiped out.

Coastal bluffs south of the capital city of Mogadishu were denuded of vegetation because of over-grazing by sheep, goats and cattle. Some 5,000 sq km of shifting dunes were created. To date only around 100 sq km have been stabilized by replanting the slopes with prickly pear cactus.

In addition to the effects of upland degradation on coasts, the direct exploitation of coastal resources – for food, fuelwood, fodder and building materials – has resulted in the loss of habitats and degradation of ecosystems.

Coral reefs have been destroyed in virtually every country of the region. Surveys carried out in the mid-1980s revealed that many reefs had been obliterated by blast fishing and the use of coral for building materials and as tourist trophies. In Tanzania, of eight reefs that were recommended for marine parks in 1968, only two could still be found in 1983. The rest had disappeared from the map, blown to oblivion by dynamite fishermen and removed by coral collectors. Mozambique's few reefs have been over-exploited by fishermen and many corals killed off by siltation.

Perhaps the greatest destruction has occurred on the islands. Comoros, Mauritius, and the Seychelles all have extensive fringing coral reefs; the remote, far-western part of the Seychelles consists entirely of coral islands and atolls. Coral sands and reefs have been excavated for cement production and building materials on all three island nations. The reefs of the Comoros and Mauritius have been pillaged by tourists and collectors. In some areas, the sea has devoured beaches and coastal lands because the protective coral barrier was removed.

Unfettered tourism has probably had the most negative impact on the coral reefs of Mauritius and Réunion. Tourist resorts, built in coralline areas, have provided reef access to thousands of trophy-hunting tourists, who think nothing of tearing apart live coral looking for shells. At low tide, unthinking tourists tramp across the coral heads, crushing them, in their search for souvenirs. Many visitors to reefs don't even realize the coral is alive, so they see no reason why it shouldn't be collected. There may be more coral from these islands on European and North American coffee-tables than remains in their coastal waters.

Fortunately, the most spectacular reefs in the Seychelles are all so remote from the main, inhabited islands that they are free from exploitation. The magnificent raised coral atoll of Aldabra was made a World Heritage Site in 1983, principally because it is one of the last refuges of the giant land tortoise (*Geochelone gigantea*), similar to the species living on the Galapagos Islands, off the coast of Ecuador. The atoll is also a refuge for thousands of nesting water birds. Coral reefs around the main island of Mahe, however, have been damaged by trophy collectors and covered with sediment from poorly-planned coastal development projects – many of them tourist resorts.

Mangrove forests in East Africa are not extensive; under a million hectares remain in the entire region. The most extensive growth – 320,000 hectares along the north-west coast of Madagascar – supports a thriving shrimp fishery. With few exceptions mangroves are endangered by coastal development, fuelwood collectors and siltation. In Kenya there are large areas of mangroves around the Tana River Delta and north of Lamu, with small pockets found at the mouth of creeks and estuaries. On the south coast there are significant stands at Gazi and Funzi south to Vanga. Tanzania has one major stand in the Rufiji Delta, plus smaller pockets scattered about, and Mozambique's 67,000 hectares are in its central delta region.

Without genuine, active conservation programmes in place to preserve what remains of the region's mangroves and coral reefs, exploitation will continue. By the time some areas are set aside, their value as reserves may be negligible.

Coastal pollution

Most of the municipal and industrial wastes which find their way into the Indian Ocean are untreated. Sewage systems in the region serve a maximum of twenty per cent of the entire population. Some claim even this figure is too high. Even though Kenya has 150 sewage treatment plants throughout the country, less than half the population is covered by them. The sewage treatment plant in Mombasa is constantly over-loaded. Dar es Salaam's sewage treatment network has completely collapsed, necessitating the establishment of a master sewerage plan for the city. To be completed in stages, the system will not be ready until the year 2010. Meanwhile, everything is dumped raw into shallow coastal waters. Somalia has no sewage network at all; relying instead on pit latrines and septic tanks, which pollute groundwater. Only ten per cent of the urban population of Mozambique had access to sanitation facilities by the mid-1980s. A cholera epidemic in Maputo was linked to the sewage-contaminated bathing waters in the Bay of Maputo. Of the island states, Madagascar and the Comoros have no sewage treatment plants. Wastes from both islands are channelled into coastal waters.

More serious, perhaps, is the amount of industrial and agricultural poisons brought into the sea from rivers and streams. All of Nairobi's municipal and industrial wastes are flushed into the Athi River, which exudes such a foul odour that the stench can be detected for kilometres. Some of these poisons eventually reach the sea. Their effects on marine communities are virtually unknown.

Industries in coastal areas dump their wastes directly into the sea untreated, a pattern that show no signs of improvement. Mogadishu's coastal waters are patrolled by huge sharks, attracted to the raw wastes from a slaughter house, which are simply drained into near-shore waters. Unwary swimmers have been snatched by voracious sharks that have even learned to swim on their sides, seizing their victims in waist-deep water.

Coastal industries in Tanzania dump untreated wastes from soap factories, sisal and sugar mills, and plastics, wood-processing and super-phosphate plants into coastal waters. At Dar es Salaam, industrial effluents have turned the waters of the Mzimbazi Creek – which flows through a mangrove swamp – into an anaerobic, stinking cesspool bubbling with sulphide gases.

In Maputo, wastes from the city's cotton and textile factories drain directly into Maputo Bay, parts of which are eutrophic and devoid of marine life.

Another problem which affects coastal waters is pesticide runoff from agricultural lands. Pesticides banned in the West, such as DDT, are still used extensively in East Africa. Residues from these pesticides seep into rivers and coastal waters. Madagascar alone sprays its agricultural fields with 20,000 tonnes of pesticides a year, most of it DDT. Shellfish beds, tainted with pesticide poisons, had to be destroyed in Madagascar and Mozambique.

Solid wastes are often dumped into coastal lagoons and mangrove swamps. On the main island of Mahe in the Seychelles, solid wastes are dumped into a lagoon, known as "stinking corner". Next to this eyesore and health menace is a marine park. Efforts to clean it up have failed, as the authorities don't know where else they can dump their municipal garbage.

As cities and towns expand, treating and disposing of municipal and industrial wastes will become major environmental issues. So far, little has been done in the East African region to tackle either the causes or effects of unsupervised waste disposal.

Oil pollution

East Africa lies on the other half of the world's major oil highway – the branch which flows out of the Arabian Gulf west to markets in Europe and North America. Approximately 475 million tonnes of oil are transported through the region every year. About half of it is carried by Very Large Crude Carriers (VLCCs), averaging about 200,000 dead weight tonnes apiece. On any given day there are about two hundred oil tankers in East African waters, going to and from the Gulf.

With this much oil being transported through the area, oil spills are inevitable, but so far no catastrophic accidents have occurred. Most of the oil pollution is due to the routine cleaning of tanks, and the discharge of dirty ballast waters and bilge slops. Captain James Ferrari, former Principal Secretary in the Ministry of Transport for the Seychelles, estimates that, on average, about 35,000 tonnes of oil are spilled into the eastern Indian Ocean every year through normal shipping operations. The actual figure could be considerably higher.

Despite the dangers, the countries of East Africa have not yet developed contingency plans for combating oil spills. And even if they had a plan of action, there is practically no equipment to use in such emergencies.

Fisheries

Most fish are landed by small-scale commercial and artisanal fishermen. For the most part, fishing takes place in shallow lagoons and on reefs. On the mainland many fishermen simply wade out to the reef, and encircle fish with simple hand-nets. Those with boats – usually outrigger canoes – stay within five kilometres of shore near coral reefs and mangrove estuaries. On the Comoros, some 4,000 outrigger canoes fish within a couple of kilometres of the coast, using hand-lines. Madagascar's 4,000 artisanal boats, mostly outriggers, are found on the calmer, western side of the island; they use hooks and lines and gill-nets. Trawling is limited to a few offshore areas, where tuna, sardines, anchovies, Indian mackerel and billfishes are caught.

In the early 1980s, fish catches in East African waters, including the islands, amounted to no more than 170,000 tonnes. Most of the harvest was taken by artisanal fishermen.

There is one exception. Shrimping is big business in Madagascar, where some 40 trawlers haul in $50 million worth of the crustaceans a year. On the continent shrimp catches are confined to a few important mangrove areas in Kenya, Tanzania and Mozambique.

Fishing is not highly developed, mainly because access to better techniques and fishing gear is severely limited. Most Kenyan and Tanzanian fishermen, for example, are confined to coral lagoons where they fish on foot during low tide. Because nylon nets are expensive, they mend their nets constantly and use them until they fall apart. Throughout the region, cold storage facilities are lacking and road networks primitive. Fish and shellfish must be sold the same day in village markets or to middlemen who pack the fish in ice and deliver them to buyers in nearby towns and cities.

The other factor limiting fish consumption is local tastes. Somalians, with their nomadic tradition, consume only 0.7 kilogrammes of fish per person a year, while the people of the Seychelles rely on the sea for most of their animal protein; the average Seychellois consumes 70 kilogrammes of fish a year. With transport networks already over-taxed on the continent, the demand for seafood is likely to remain coastal.

Despite all the evidence against the non-sustainable harvesting of fishery resources, over-fishing is becoming a serious problem along much of the coast of Kenya and Tanzania. As increasing numbers of people are pushed off upland farms, many migrate to the coast in search of employment. Even though they use primitive gear, so many people are fishing that fish stocks are becoming depleted in lagoonal areas. Inappropriate fishing practices abound as well. Many of the newcomers, lacking traditional knowledge of the sea and its creatures, harvest fish and crustaceans bulging with eggs, or take fish before they have matured. Others use destructive equipment, such as overly fine-meshed nets which collect everything, even fish eggs. Whether out of greed, ignorance or desperation, increasing numbers of subsistence fishermen are resorting to dynamite, particulary in Tanzania, which not only kills indiscriminately but destroys the reef upon which their livelihoods depend.

The two sides of Mombasa

Akida Abushiri's greatest fear is that he will have to give up the sea. He has been fishing in the clear coralline waters north of Mombasa for over 45 years. But his fishing days may be coming to an end. Over the past decade, many people, unable to find employment on the land, have turned to the sea for a living. This has meant that edible fish are now harder to find and catch.

Akida lives in the village of Shariani, in a small traditional hut made from mud and wattle, with a palm-thatched roof, called *nyumba ya kiswahili* in Swahili. There is no running water and no sanitation facilities. Pit latrines are dug whenever necessary and fresh water is taken from shallow wells or bought from vendors. He has lived on this same plot of land for 50 years. Like all of his neighbours, however, he does not own the land; the government has allowed the people of Shariani to squat here because its sandy soil, filled with pieces of ancient coral reefs, is poor for agriculture.

It is an unsettling existence. "The government could always come in here and tell us to move somewhere else," points out one of Akida's relatives. "We have nothing to fall back onto, but the sea."

"The sea," insists Akida, "is just as productive as before. But now more people are fishing, so there is less to go around. When I started fishing in this area 45 years ago, there were only five other fishermen. Today, about 40 people fish regularly in this part of the reef."

Increased numbers of fishermen are only part of the problem facing Akida and his neighbours. The poor fishermen of Shariani have no boats, not even simple dugouts. Without boats, they cannot fish outside the reef, where fish are more plentiful. Like most of the other older fishermen from the village, Akida uses a simple net, made from nylon or cotton fibre, to catch fish – a technique that has been used along this part of the Kenyan coast for decades. With two of his comrades, Akida wades out to the reef at low tide. When they find a likely spot, usually up to their necks in water, the net is spread out and readied. While one fishermen beats the water with his hands, the other two guide frightened fish towards their net. Any fish taken are quickly placed in fish bags, woven from the leaves of the screw pine (a large plant). There are no cold storage facilities in the village, so fish must be sold the same day in the local market, or to fish vendors from nearby villages.

At the best of times it is back-breaking, dangerous work. Wading out to the reef, fishermen have to take care not to step on poisonous sting rays and stone fish. Once near the reef, they must be on the lookout for big waves which can crash over them unexpectedly, pounding them into the jagged edges of the coral. During the first half of 1989, two fishermen from this village were killed on the reef, both victims of freak waves which crushed them against the coral. Their bodies were not recovered for days. Anyone who survives being dragged over the coral reef faces treacherous surface currents which can sweep someone far offshore in minutes.

Like many other fishermen in Shariani, Akida must support his family on a meagre income. He averages about 1,000 Kenyan shillings a month ($55). From this he supports two wives and ten children. Four children have already left home and have families of their own. Since his land is too poor to grow vegetables, maize, rice and other staples must be bought.

The good times are not as good as they once were, and the lean times are leaner. "Twenty years ago, I could save some money in the bank," says Akida, "but not today. One good week is usually followed by several bad ones. So any surplus is quickly used up."

Sometimes it is difficult for Akida and his neighbours to afford new nets. They are expensive – 1,300 shillings each – and must be bought in Mombasa, a 45-minute bus ride to the south. "We should buy new nets twice a year," points out Akida, "sometimes we just don't have the money."

Although there are no reserves or protected areas in this part of the fringing reef, which stretches for several hundred kilometres along the Kenyan coast from Mombasa northwards, the older fishermen of Shariani and other villages are natural conservationists. They don't use destructive fishing gear, like dynamite and poisons. They are careful not to harvest fish with eggs, and when net fishing they treat the coral as if it were rare porcelain.

Many of the newcomers to the coast are not so careful about what techniques they use to catch fish. Some of the younger fishermen are using "juya nets", with very small holes. "This is a dangerous net, since it takes everything, even fish eggs," explains Akida.

The other scourge afflicting Akida's village is spear-fishermen. Within the last several years, Shariani and other nearby fishing villages have seen an influx of "school leavers". Unable to find suitable employment, they have taken to the sea to make money by spear-fishing at night and collecting corals and shells for the tourist trade during the day.

Akida claims they are interfering with his subsistence fishing by taking everything they can get. "The government should clamp down on this form of fishing. Spear-fishermen spare nothing and they damage the coral when they miss their mark."

Twenty-six year old Jackson is one of the spear-fishermen Akida is complaining about. He left school and could not find anything worthwhile to do, so he has turned to spear-fishing on the reef at night using a water-proof flashlight to blind the fish.

He justifies his methods because they work. "We can get 20 kilos of fish a night," says Jackson. "We only go out for about three hours a night. Day fishing is too much work and there are too many guys doing it."

Usually, Jackson fishes with nine other young spear-fishermen. They all use crudely-fashioned home-made spear-guns. "But night fishing is really easy," he adds. "The fish are resting and don't expect to be taken at night. You can swim right up to them and almost grab them with your hands."

Jackson and his friends each make about 4,000 shillings a month from night fishing, four times the amount Akida brings in. And they do not have families to support.

Although Jackson insists that he does not take shells for the tourist trade, many young men in the area do. Because trochus and triton shells have been pillaged, the reef is suffering from plagues of the crown-of-thorns starfish and sea urchins. The natural balance has been upset.

With all the new fishing pressures to contend with, Akida wonders how he will manage a year from now. Sitting amidst his children on a palm mat, Akida meticulously repairs his only net. The strain of the last few years shows on his wrinkled brow.

The government has overlooked the north coast of Mombasa. None of the fishing communities here has received training in new techniques, or loans to buy better equipment. Part of the reason lies in their own stubbornness. Distrustful of organizations, they have not been able to form co-operative fishing associations, which might attract government project support, or international aid. An attempt by one young man from the village to form a co-operative came to nothing. The fishermen did not want to put in any of their own money.

"All we really need is a small boat and two new nets," observes Akida. "Then we could fish outside the reef and make a better living."

Until that day comes – if it does – he must catch what he can in the lagoon, and watch as the waters on the other side of the reef fill with fishing boats taking in rock cod, barracuda, groupers, mullet and billfish. And he must watch as younger fishermen, less mindful of the ecology of the reef, take more fish and shellfish than they should. For the poor people of Shariani there is only less of everything.

By contrast, on the south side of Mombasa fishing villages have prospered by forming co-operatives. Mohammed Kitawana is one of the officers of the Mwagugu Fisheries Co-operative Society, set up ten years ago in the fishing village of Vanga, a few kilometres from the Tanzanian border. The extensive growth of mangroves which separates the village from the sea provides a rich and diverse catch including shrimp, prawns, rabbitfish, squid, mullet and many other off-shore species.

Nearly all of the fishermen of Vanga are members of the co-operative. Kitawana counts 179 members and more people are writing to join every month. Unlike Shariani on the north coast, Vanga's fishermen all own their own boats and dugouts. They use a variety of techniques including hooks and lines and different kinds of nets, depending on what type of fish they are after.

"Some of our fishermen specialize in shrimp and lobsters," explains Kitawana. "Others go after mullet and rabbitfish. But most of the fishermen here harvest whatever edible fish they can catch."

Mohammed Kitawana is thirty years old and has a high school education. He was born in Vanga and is happy to be able to contribute to the advancement of his village. He is clearly pleased at the success of the co-operative.

"It took us two to three years to set it up in the beginning," confesses Kitawana. "We had to get government permission and organize all the fishermen. Some were sceptical at first that we could make this work. But today, the co-operative's success is well-known. All our members make a comfortable living from the sea."

The co-operative sells no less than 500 kg of fish and crustaceans a day and often unloads up to 1,000 kg. "On a monthly basis we probably take in 100,000 shillings," says Kitawana. "After all expenses are paid, including the fishermen's share of the profits, we still have around 10,000 shillings left in our reserve fund."

A few years after the fishermen's co-operative was established, the government built the Vanga Fish Depot, where the co-operative sells its fish. They have also built an office. "Once you have an organization, you can get bank loans and assistance from the government," points out Kitawana.

There is no over-harvesting of fish around Vanga and none of the fishermen resorts to illegal techniques, like dynamite fishing. Occasionally, Tanzanian poachers infringe on Kenyan waters, but anyone caught dynamiting is quickly driven off. "If anything, our waters are still under-fished because we don't have big trawlers in these waters," notes Kitawana.

Some of the fishing villages south of Mombasa stand in sharp contrast to the poor, struggling communities north of the city. One of the reasons for the dichotomy is because the mangrove-based fishery here is more productive. But the key ingredient, missing from the north coast, is a fisherman's organization.

"It's not the answer to everything, but forming a co-operative or some kind of organization that represents your interests is one way to protect yourself against outside forces," says one fishermen from Vanga. "We no longer feel that we are out there alone."

Marine conservation

With so many other problems to worry about, government officials in the region have not paid much attention to the conservation of marine areas. Kenya and Tanzania – currently engaged in an escalating bush war with poachers – are more concerned about saving what they can of their terrestrial animal life. So far, there are only 41 protected marine areas in the entire region, including the islands. And more than half of these are in one country: the Seychelles.

Kenya has designated four marine parks, covering 50,000 hectares, but most of the area protected is coastal land, not near-shore waters. Somalia and Madagascar have no marine parks or protected areas. Tanzania has recently gazetted seven marine reserves. Mozambique has designated two marine reserves, comprising 8,000 hectares; but management is lax. Mauritius has managed to conserve a few marine areas, mostly as tourist attractions. The Seychelles have an impressive record of marine conservation, setting aside 38,000 hectares of coral atolls and islands, with more areas under consideration.

The status of marine mammals is uncertain. Dugongs are extremely rare and those remaining are threatened by poachers and loss of habitat. Sea turtle populations have also been devastated by poaching and destruction of habitats, despite national legislation against harvesting. Dolphins, on the other hand, seem to be in no immediate danger in East African waters.

On the bright side, the Indian Ocean was declared a Whale Sanctuary in 1983 and has remained one since then. No commercial whaling activities are permitted at all in the entire ocean basin, north of the 55° south parallel. Although this caveat exempts Antarctic waters, where many species of whales spend their summers feeding on krill and plankton, it is still a significant breakthrough in global efforts to conserve these vanishing mammals.

East African Action Plan

In 1985 the countries of the region banded together to form the East African Action Plan, under the auspices of UNEP. The first phase, now being implemented, is the formation of scientific contact points within the region. Participating institutes and research centres are working together in efforts to make their methodologies, and results, compatible. Baseline pollution studies are being carried out, in order to pinpoint local problem areas and assign priorities.

The Action Plan calls for the following measures:

- an inventory of ecosystems in the region;
- regional inventory and status report of all existing protected areas;
- documentation of all endangered species;

- facilities made available to train pollution control technicians;
- assessment of socio-economic activities that may have an impact on coastal and marine environments;
- strengthening of oil spill contingency plans;
- studies of ciguatera poisoning, and heavy metal and organochlorine contamination of fish and shellfish; *and*
- assessment of the effects of dynamite fishing on coastal habitats and fish stocks.

It has been a slow process, but building institutions takes time. The mainland and island states of East Africa started the programme with few environmental experts and little in the way of institutional support. Building up a research network of institutions and experts consumed the first few years. To this end, UNEP has contributed nearly $1 million to get the programme started. Now that the Action Plan is moving out of its "study phase", the region may begin to see some action designed to solve the problems confronting coastlines.

Many needs have not been met. Public awareness campaigns are critical to the overall success of coastal management plans. But so far, not one African NGO deals with coastal issues and there is little government interest in coasts, other than for agriculture and tourist development.

Resource-use planning must begin to be implemented along the coasts, at least in a few of the priority areas as identified by the Action Plan. "Right now we are doing little more than crisis management," observes one Kenyan biologist. "We need to develop the capacity to plan for people and their needs."

In this context, controlling urban growth must be a top priority. A related issue is dealing directly with the causes and consequences of landlessness. Ways must be found to keep farmers productive and on their land, and employment must be found for urban squatters and slum-dwellers. When the causes of poverty are addressed directly, then coastal-zone problems, which are themselves related to people's poverty, will be easier to resolve.

11. West and Central Africa

This vast region includes 21 coastal states, stretching from Mauritania in the north to Namibia in the south. The economies of the region vary considerably; Nigeria is a rapidly emerging industrial power, while the Gambia and Benin are almost subsistence economies.

Despite the economic, political and social divides separating West Africa, the countries of this region have many resource and environmental problems in common. Nearly all of the countries of West and Central Africa must fight coastal erosion, the most severe environmental problem of the region. Although partly a natural process, the sea is eroding more of the coastline than normal because development along the coast has upset the natural balance between land and sea. Coastal towns and cities sprawl in all directions as their populations continue to expand. As in East Africa, most urban and industrial wastes enter the sea without any kind of treatment. Tropical forests have been decimated and mangroves cleared. Oil pollution is widespread, especially around the coasts of oil-producing nations like Nigeria. And rivers and streams dump toxic residues from pesticides and fertilizers into coastal waters.

For most of these countries, coastal management is barely considered, let alone enacted. Meanwhile, development along this long and varied coastline seems to be divorced from any constraints.

Ironically, one area of "development" has given impetus to the regional approach to environmental protection: the importation of hazardous wastes. With some countries willing to accept the toxic wastes of the First World and others bitterly opposed, the need for a clear set of guidelines and rules governing the transport, storage, treatment and disposal of such wastes became obvious. The Action Plan for the West and Central African region (called WACAF by UNEP) has, on this issue, become a medium of central importance.

Geography of a continent

The 21 countries of West and Central Africa cover 9 million sq km, or nearly thirty per cent of the entire continent. Their collective coastline stretches out over 6,500 km. Coastal shelves extend out only 30–40 km offshore, then slide down to ocean basins. But even the shelf, which averages 100 m in depth, is fractured by deep trenches. Coastal upwellings make the continental shelf a rich fishing ground.

Excluding Arab North Africa, the countries of this region are the most developed on the continent. Nigeria has built its economy on oil and gas, while countries like Senegal, Zaire and Liberia have evolved advanced mining and processing industries. Only a few West African countries are on the United Nations list of the least developed. Most of these nations have an abundance of natural wealth to build on: minerals and hydrocarbons, timber, rich agricultural land and fisheries.

But the real test here, as over much of the rest of Africa, is whether they can continue to develop without ruining their resource base in the process. The task is formidable. Population growth is already making a mockery out of government attempts to manage resources and the environment. Cities are growing especially quickly; and in many parts of the region the coasts are suffering the most from the rapid rise in population.

Population growth

The average rate of growth for the entire region is close to three per cent a year, one of the highest in the world, doubling human numbers every 23 years.

As of the late 1980s, the countries of West and Central Africa contained nearly 200 million people. At least a quarter of these are thought to reside in coastal areas. Indeed, with a few exceptions, such as Kinshasa, Zaire, the capitals and major industrial areas are all coastal.

Coastal capitals and other urban areas are growing at more rapid rates than towns in the interior. In some cases, such as Lagos, Nigeria, land is at a premium. In 1985, for example, Lagos was ranked as the world's twenty-fourth largest city with an official population of 6.1 million. But in terms of population density, the city came in second place after Hong Kong, with nearly

42,000 people per sq km. The city's growth rate is already outstripping public services. Planners pale at predictions that Lagos is expected to have ten million residents by the turn of the century – although some claim this figure has been reached already. Much of the city's expansion is taking place around its sheltered lagoon.

In efforts to stem population growth, many West African governments have instituted family planning programmes. However, since most West African cultures view big families as a form of social security, little progress has been made in bringing down birth rates.

Major urban centres in West Africa, as elsewhere, also concentrate manpower, skills, industries and commerce. The capital of the Ivory Coast, Abidjan, contains 20 per cent of the population of the country and accounts for more than 70 per cent of all economic and commercial transactions. Similarly, Lagos – with a tenth of Nigeria's population – accounts for nearly 60 per cent of total value added in manufacturing and has 40 per cent of the nation's highly-skilled labour force. Such trends are region-wide. The effects of urbanization and the concentration of industrial and commercial activities along coasts has not been fully taken into account by government planners. If unchecked, coastal resources could be critically depleted by this centralization of activity.

Coastal pollution

As on the east coast, West Africa pumps most of its municipal and industrial effluents directly into coastal waters without any form of pre-treatment. Raw sewage from roughly fifty million people is channelled into shallow coastal waters, bays and lagoons. In some areas high concentrations of bacteria pose a clear threat to human health. Lagoonal waters around Lagos receive more than sixty million litres of sewage water a year. Typhoid fever and infectious hepatitis are rampant, caused by the consumption of shellfish tainted with sewage.

The coastal waters of virtually every major urban centre are clogged with a vile assortment of municipal and industrial wastes – everything from sewage to heavy metals and chemical poisons. West African waters are more severely affected by industrial wastes, perhaps, than any other part of the continent. Over sixty per cent of all industries in the region are coastal. Untreated effluents from petrochemical refineries, mining and metal smelting operations, and food processing, chemical and textile plants all discharge their wastes into coastal waters.

The cement industry in Gabon and Togo contaminates near-shore waters with suspended solids, turning estuaries a pallid grey. Wastes from textile manufacturing in Ghana, Benin and Nigeria go directly into coastal lagoons, creating locally high concentrations of phenols and chrome. Heavy metals from mining and metal smelting ruin shellfish beds along the coasts of Liberia, Senegal, Guinea and Sierra Leone.

The other affliction of river basins and estuaries is the great quantities of pesticides sprayed on agricultural lands throughout West Africa. Since agricultural exports are still the basis of many of the region's economies, West and Central African states consume around ten per cent of all pesticides produced in the world every year. "Practically all the fish, crabs, and shrimp in the coastal areas are exposed to pesticides," states one UNEP report on the coastal health of the region. Once pesticides infiltrate coastal ecosystems, they tend to accumulate in the fatty tissues of marine organisms. High levels have been detected in crustaceans and shellfish. The effects on human health of pesticide accumulation in marine life have not yet been studied sufficiently.

"One of the top priorities of our Action Plan is to control the discharge of municipal, industrial, and agricultural wastes into the coastal waters of the region," points out Dr A. Chidi Ibe, Head of the Physical and Chemical Oceanography section of the Nigerian Institute for Oceanography and Marine Research in Lagos. "But so far we are still in the evaluation stage."

Oil pollution

With most oil in the region being pumped from offshore platforms in the Gulf of Guinea, particularly off the coast of Nigeria, it is not surprising that the coastal waters of the entire Gulf are contaminated with oily wastes and tar balls. The worst oil spill occured in Nigerian waters, when an offshore oil well blew out; some 421,000 barrels were sprayed into the Gulf, killing mangrove forests and fouling beaches.

The entire delta of the Niger River – Africa's third largest, which drains a huge section of West Africa – is polluted with oil residues from offshore drilling and transport.

Oil reserves in West Africa total 2.4 billion tons with 80 per cent of it belonging to Nigeria. At current rates of extraction, these reserves will last another two decades, perhaps longer. Oil-spill contingency planning, although still unfinished, is high on the list of priorities for the regional action plan.

Forest loss

West African tropical forests and mangrove swamps have been devastated in the name of development. The survival of the region's forests beyond another 50–100 years is doubtful. Forests throughout West Africa are being cleared for urban expansion and the production of food and cash crops, and exploited for timber and fuelwood. Both Nigeria and the Ivory Coast are losing over five per cent of their forests every year.

Scientists fear the loss of tree cover will mean less rainfall and increasing temperatures in some areas. If such delicate balances are upset, the consequences for agricultural production could be catastrophic.

At the same time, mangroves are being cleared to make room for the expansion of towns and cities, industrial estates, and fish and shrimp ponds. Now, the last major stand of mangroves in West Africa – the 900,000 hectares at the delta of the Niger River in Nigeria – is threatened. Seaward, these mangroves are victims of chronic oil pollution and beset by coastal erosion. On land they are exploited for timber, tannin, fuelwood, fodder and fish ponds.

Coastal erosion

"Coastal erosion is probably the most serious environmental problem facing West Africa," says Dr Ibe. The reasons are due to a complex assortment of ocean–land interactions. But in plain terms the ocean here acts like a huge scoop, gouging the land and depositing the debris in deep trenches which knife across the continental shelf fairly close to shore.

The effects are obvious. Hundreds of coastal villages have been moved further inland, as the sea reclaimed their land. In the Gulf of Guinea river deltas had helped build new land by depositing silt from the watershed. Many of these sand spits were occupied by small fishing communities who now have to evacuate their islands as the sea advances.

Up and down the coast, agricultural lands are being washed out to sea. The twisted trunks of palm trees and mangroves sticking up out of silt-laden coastal waters are a common sight. Sand banks designed to hold back the sea have been battered down. Coastal cities have been forced to reinforce tidal barriers and jetties.

Coastal erosion has reached such proportions in land-short countries like Togo and Benin that they have asked the French Ministry for Co-operation for help in combating it. Off the western pier of Lomé Harbour (Togo), the sea has advanced 380 m in 15 years. The entire eastern portion of the harbour itself is being lost at the rate of 20 m a year.

The great mangrove forest of the Niger Delta are being pushed back towards land by the sea. Scientists estimate they are losing ground at rates reaching tens of metres a year. The huge amounts of sediment transported into the delta by the river is swept away by the currents.

To what extent human interference has contributed to the process of coastal erosion is unknown. But unplanned coastal development is most certainly a factor in some areas. "This problem is receiving a great deal of attention by scientists in the region," asserts Dr Ibe. Unfortunately, the costs of doing something about it are often prohibitive.

Fisheries

Despite a relatively narrow continental shelf, upwellings of nutrient-rich waters are found in certain areas along West Africa's coastline. The region has no fringing coral reefs, but mangrove-based fisheries abound in the north-central regions. The cool Benguela Current which flows north off the coasts of Namibia and Angola supports an annual catch of 1.5 million tonnes (mostly pilchards). Altogether, about 3.5 million tonnes of fish and shellfish are caught in the region every year. The largest catches are taken by Namibia, followed by Nigeria, Senegal, Ghana and Angola.

Fishing is an important source of revenue for West and Central African states. Mauritania's exports of seafood rank second to iron ore. Senegal is the region's leading producer of shrimp, which are taken by trawlers in the mangrove delta of the Saloum and Sine Rivers. About seventy per cent of the catch is exported. The Gambia has developed an export industry based on crabs, oysters and lobsters.

Liberia, Ghana, Togo and Benin have thousands of fishing communities. Most fishing is confined to near-shore waters. In

Ghana, fish consumption exceeds that of meat and the industry is credited with about 1.2 per cent of the country's total GNP.

Nigeria's fisheries are also lucrative, amounting to around 200,000 tonnes a year. Much of the catch is taken by artisanal fishermen with motorized dugouts using simple hook-and-line techniques and gill nets. Still, demand is so great that three-quarters of Nigeria's fish is imported.

Large processing ships from the Soviet Union, Spain, Japan and Norway trawl off the continental shelf. These fleets can catch, process and freeze great quantities of fish before returning to shore. Most tuna taken in West African waters are caught by these foreign "process fleets".

The primitive state of the fishing industry in many West African countries means they do not get a large slice of the offshore resource. The UN Food and Agriculture Organization has long been active in West Africa in efforts to help upgrade fishermen's skills and equipment. FAO reckons that the commercial-fishing potential of the region is immense, amounting to perhaps six million tonnes a year.

Hazardous wastes

The West African region clearly has been targeted by international waste-handling firms and chemical companies as a possible repository for a frightening variety of hazardous industrial and municipal wastes. In 1988 the small coastal town of Koko in Nigeria made headlines when local residents suddenly developed unexplained skin rashes and suffered bouts of vomiting, diarrhoea and headaches. The symptoms were traced to a dump site where investigators discovered that 4,000 tonnes of toxic chemical wastes, left there by an Italian waste company, were leaking poisons into the air and water. Outraged, the Nigerian government impounded Italian ships and arrested Italian businessmen, until the Italian government agreed to remove the dangerous wastes. But the final chapter was a grisly one: Nigerian dock workers who loaded the wastes onto an Italian ship did not have protective masks; many of them began coughing up blood.

Greenpeace International, which has been monitoring the shipment of hazardous wastes from the developed North to the developing South, claims that such waste shipments are increasing in frequency. Between 1986 and 1988, say Greenpeace monitors, industrialized countries have shipped around 3.7 million tonnes of all kinds of wastes to the developing world. Much is thought to be highly toxic PCBs, pesticide residues, solvents and incinerator ash.

A Greenpeace review of shipments, or attempted shipments, to West Africa include the following examples:

- Guinea-Bissau recently cancelled a five-year contract to take 15 million tonnes of pharmaceutical and tanning wastes from Europe.
- When Guinea discovered that 15,000 tonnes of toxic flyash had been dumped on Kassa Island by an American waste firm, the government filed an official protest with the State Department and the wastes were eventually returned to sender.
- Sierra Leone has accepted toxic incinerator ash from the US containing cadmium and mercury.
- The government of Benin has agreed to take 5 million tonnes of industrial waste each year from the US and Europe.
- Liberia is considering offers to take hazardous wastes, including contaminated earth, from West Germany.
- Gabon has agreed to accept radioactive waste from uranium mining in Colorado, USA.
- The Congo cancelled a deal to accept one million tonnes of chemical wastes from Europe and the US.
- Angola has agreed to dispose of unspecified quantities of toxic waste from Europe.

West Africa seems to be a popular target for such arrangements because the governments are thought to be cash-short and unconcerned about the health of their environment. Nigeria's reaction has ended that myth. But the region is still bombarded with proposals to take toxic wastes off the hands of Northern producers. It is an issue that will certainly require a regional approach for its resolution.

The Action Plan for West and Central Africa

The Action Plan for WACAF contains three main areas: pollution control; fighting coastal erosion; and developing a contingency plan for combating oil spills.

Although the WACAF countries met for the first time in 1981 to discuss the establishment of a West African Action Plan for protecting and managing their coastal environments, the Abidjan

Convention for Cooperation in the Protection and Development of the Marine and Coastal Environment of the West and Central African Region did not come into force until 1984.

Meanwhile a number of important baseline studies were carried out in an effort to determine what major environmental problems confront the region and to make recommendations for their control. Now pollution monitoring is being carried out throughout West Africa by a network of interconnected research institutes. Masses of data have been collected on the region's major problem areas. A Trust Fund was also established to pay for the programme. As of 1989 it contained only $387,000, but more money is expected.

Still, with so little funding it is hard to see how WACAF can achieve much in their fight to conserve resources and protect their eroding coastlines. One West African scientist says that "the time for talking is over, we need action to deal with these problems." The epidemic of toxic waste dumping in the region may be the one issue capable of galvanizing West African governments into collective action.

12. The Red Sea and Gulf of Aden

Nine countries border the Red Sea and Gulf of Aden, but only seven are presently participating in the Regional Convention for the Conservation of the Red Sea and Gulf of Aden: Jordan, Saudi Arabia, Sudan, Somalia, the Popular Democratic Republic of Yemen (South Yemen) and the Arabic Republic of Yemen (North Yemen). The Palestine Liberation Organization (PLO), representing Palestine, is also a member.

The region, surrounded by dry and arid country, is not developed, except for Egypt, Jordan, and Saudi Arabia. Most of the countries are relatively poor, with few resources other than oil.

In this case, poverty, combined with a lack of polluting industries, has saved the Red Sea from the depredations inflicted on the Arabian Gulf. The sea is pollution-free, except around major urban centres and the offshore oil fields in the Gulf of Suez.

Divers who have visited the Red Sea claim that its coral reefs are among the most outstanding in the world. But other than Jordan and Egypt, none of the countries of the region has developed tourism. If developed with sensitivity, tourism could bring needed foreign currency to their economies.

The Red Sea as history

With the opening of the Suez Canal in 1869, the Red Sea was transformed into an important commercial crossroads: the canal provided access into the Mediterranean and European markets. Some 20,000 merchant ships and smaller tankers sail its waters, going to and from the Persian Gulf and Asia every year – a full 15 per cent of all international shipping.

With the ships have come fish as well, migrating from the Red Sea into the Mediterranean. Since the canal opened, over forty different species of Red Sea fish have taken up residence in the eastern Mediterranean. There is even a special name for these fish: Lessepsian migrants, after Ferdinand de Lesseps, the French engineer who built the canal.

The Red Sea is long, narrow and deep. The sea, one of the warmest and saltiest on earth, is 2,000 km long and averages 2,000 m in depth. There are practically no continental shelves; its shallow coastal-waters drop off steeply. Fishing is limited to coral reef areas, intertidal flats and islands.

The Red Sea, however, has inestimable value as a research laboratory for marine science. It is, as one researcher put it, "an ocean in the making". It is one of the youngest sea systems on earth and of immense interest to marine biologists and geologists. "Physical oceanographers also consider the Red Sea Basin, with its peculiar physiographic and boundary conditions, to be unique," explains Makram Gerges, Programme Officer for UNEP. "With these features – a semi-enclosed sea with deep basins and steep slopes – we consider it an excellent natural physical model for geophysical fluid-dynamic studies." Due to the semi-isolation of its bottom waters, the sea is filled with unique and endemic species of fish and plants. Because the sea's coastlines have not been developed and urban centres are few, most of the region's marine communities are intact and undisturbed.

Population

The number of people living along the shore of the Red Sea is not known. But rough estimates put it at no more than five million. Urban centres are scattered, with hundreds of kilometres of barren sand-dunes between them. Still, there are some important urban areas on the sea: Port Sudan and Jiddah in the south, and Ghardaqa and Suez in the north. Along the Gulf of Aden three more urban centres are found: Djibouti, Aden, and Al-Mukalla.

The main reason the area has been avoided by development is an acute lack of fresh water. Saudi Arabia now extracts its drinking and municipal water from the sea with the help of giant desalination plants, and Jiddah is growing and prospering as a centre for commerce and government, one of the few cities along the sea to do so. Most of the towns found along the Red Sea are little more than outposts in a sea of sand.

Resources

Resource exploitation is not a widespread problem, for the simple reason that there are few people around to exploit anything. Coral reefs are undamaged except in the oil fields of the Gulf of Suez in the north. Mangroves – pushing their northern-most limits – are troubled more by grazing camels than people. In a few areas, mangroves are a valuable source of building materials, fodder and fuelwood. But exploitation is local and limited. Only one stand of mangroves, along the southern Saudi Arabia coast, has been severely damaged. These were simply over-grazed by camels and goats.

The nine species of seagrasses found in the Red Sea are in relatively healthy condition. Coastal dredging along the Saudi and Egyptian coastlines has ruined seagrass habitat, but again, the effects are localized.

Coral reefs flourish in the Red Sea. There are over 150 species of coral in the northern and central sections of the sea. It has been estimated that the standing crop of fish on Red Sea reefs is 10–15 times as productive as the North Atlantic. Most reefs are in good condition, thanks again to low population densities and the mostly artisanal nature of Red Sea fisheries.

Metalliferous muds, rich with metals, are found on the deep seabed, midway between Saudi Arabia and the Sudan. When geologists estimated that the muds could contain two million tonnes of zinc, 500,000 tonnes of copper, 80 tonnes of gold and 4,000 tonnes of silver, these two countries formed the Saudi–Sudanese Red Sea Commission in 1974, in a bid to exploit the mineral resources of the seabed. Despite research investments of over $100 million, mining has still not commenced. The technology needed to extract valuable metals from such depths has not developed as fast, nor proved as economic, as previously thought. The Commission, however, insists that commercial mining operations should begin in the early 1990s.

Pollution

With only a handful of cities of any size along the sea, municipal and industrial wastes pose few problems to the Red Sea's ecology. Jiddah has a sewage treatment network, so wastes are not dumped raw into coastal waters. Port Sudan lacks any real treatment, but sewage is swept away by currents and dispersed. Most of the other small towns have no sewage treatment plants; wastes are dumped into coastal waters. However, the effects of municipal wastes on coastal ecosystems are thought to be minimal. Except for petrochemicals, there are few industries along the entire coastline.

The one blot on the Red Sea is oil pollution. Nearly 100 million tonnes of crude oil are transported through the Gulf of Aden and Red Sea every year. But most of the oil pollution is concentrated around the oil fields in the Gulf of Suez, in Egyptian waters. Tarballs and floating slicks are common in this part of the region. How much oil is spilled into the Gulf of Suez from offshore drilling is not known, but fouling of beaches is extensive. "Some beaches along the Gulf of Suez and its islands are already oiled beyond recovery," states Dr Youssef Halim, of the Department of Oceanography at the University of Alexandria in Egypt. Reefs in this region have been damaged from oil pollution, as have seagrass beds.

There is tarring of Saudi beaches, since prevailing winds and currents sweep some oil down the coast. Oil pollution is also present in the Gulf of Aden, as ships de-ballast and clean tanks before entering the Red Sea.

There is another potential source of pollution to the Red Sea: the possible Saudi–Sudanese exploitation of metalliferous muds on the sea bottom. If mining commences in the early 1990s, the disposal of sludges and muds from this massive operation could pose real problems, especially if the wastes are merely dumped into coastal waters or spread out over the open sea. The Red Sea Commission claims that waste disposal from the mining has been considered and no serious damage will be done to the sea's near-shore ecology. There are plans that the worst of the wastes will be landfilled in specially-prepared desert repositories.

Fisheries

Fishing in the Red Sea is mainly still a small-scale, artisanal activity. South Yemen's fishing fleet, for example, consists of 1,000 small *sambuks* with outboard motors. A beach seine fishery also exists for immature sardines and anchovy. Ethiopia's fleet, at last count, contained 130 boats, but only 22 were motorized, and more than half were not sea-worthy. Sudanese fishermen operate out of small dugouts or canoes and use hand lines to catch reef fish like rock cod, red snapper, coral trout and emperors.

Saudi Arabia's Red Sea fishery is not nearly as developed as its Arabian Gulf operations. Artisanal fishermen use small dugouts or planked canoes and fish close to shore. However, the government recently formed a company aimed principally at exploiting the shrimp fishery for export to Japan. New fishing harbours are also being built along the Saudi coastline.

Egypt operates around 75 trawlers in the Gulf of Suez, harvesting mostly horse mackerel, Indian mackerel, round herring, sardines and shrimp. In the Gulf of Aden, Somalia and North Yemen have instituted small-scale trawling operations.

According to the FAO, the total take of fish and shellfish from the Red Sea and Gulf of Aden is small – amounting to only 145,000 tonnes in 1981 (about average for the past decade). Tuna are not exploited much, but estimates of their potential annual yield have been set at 20,000 tonnes.

South Yemen has developed a deep-sea lobster fishery over the last decade and now yields about 300 tonnes of lobster tails a year for the export market. But recently, catches have fallen dramatically and government officials fear over-harvesting may be the cause. In an effort to give stocks a chance to recover, South Yemen has limited the number of lobster boats to four during the season and no fishing of egg-bearing females is permitted.

Conservation

With limited human impact in the southern part of the Red Sea, it is not surprising to learn that marine turtles are thriving there. Researchers along the Sudanese coast report the highest hawksbill turtle nesting densities ever observed. Thousands of them are also found on Saudi Arabia's deserted offshore islands.

Hawksbill and green turtles nest on deserted beaches along the coast of South Yemen. Ithmun and Sharma beaches are counted among the most important nesting sites in the world for green turtles.

Exploitation of turtles is very low, and there are plenty of good nesting beaches along the southern part of the Red Sea and the Gulf of Aden. Still, no conservation legislation exists to protect these important nesting sites.

Parks and protected areas are now being set up in the region. For example, Egypt has designated the exquisite coral reefs of Raas Mohammed, at the lower tip of the Sinai Peninsula, as a protected area. Surveys are being carried out in efforts to establish a network of protected parks and reserves in other countries as well. It is hoped that Sudan's first marine national park will be set up on the Sanganeb atoll, northeast of Port Sudan. Saudi Arabia has undertaken a survey of important shallow-water marine habitats in co-operation with IUCN, the Red Sea and Gulf of Aden Environment Programme (PERSGA), and UNEP. As the Action Plan evolves, more marine conservation work will be possible.

The Red Sea and Gulf of Aden Environment Programme (PERSGA)

After several preliminary meetings the governments of the region met in Jiddah in January 1981 to approve the Action Plan for the Conservation of the Marine Environment and Coastal Areas of the Red Sea and the Gulf of Aden. The Regional Convention was adopted a year later by six of the region's states, along with a protocol to combat pollution from oil spills and other harmful substances.

Since many of these countries were starting virtually without any relevant institutional contact points, one of the first priorities was getting the member countries to set up a network of scientific institutions capable of co-ordinating the work. A Secretariat for PERSEA was established in Jiddah, under the auspices of the Arab League Educational, Cultural and Scientific Organization (ALESCO).

Building institutional support for the Action Plan was not easy. South Yemen, for example, did not have a marine research centre until 1983, when one was built with assistance from UNESCO and the Islamic Development Bank. North Yemen still does not have a functional marine laboratory, but intends to establish one.

Meanwhile, as part of the Action Plan, Djibouti, Somalia and South Yemen have agreed to establish a "sub-regional response centre" for combating oil spill emergencies in the Gulf. The centre is being set up with financial assistance from IMO and other donors.

ALESCO will also convene intergovernmental expert groups to draw up additional protocols aimed at controlling pollution from land-based sources, limiting pollution from the exploration and exploitation of the seabed, forging scientific and technical co-operation in the field of marine science, and promoting the sustainable utilization of the marine environment.

So far, baseline studies of pollution sources to the marine environment have been carried out and inventories of marine resources completed. There is a wealth of basic oceanographic data on the Red Sea; Egypt has been studying the physical and chemical properties of the sea since the 1930s. As more data accumulate, the countries of the region will be better prepared to implement coastal zone management plans. More through fortune than planning, the Red Sea has been spared from much of the coastal pollution inflicted on other seas. Hopefully, the Action Plan will help the countries of the region to keep it that way.

13. What Future for Regional Seas?

With coastal resources over-exploited and more people pressing along the coastlines in the developing world, we are courting disaster on a monumental scale. Although some fifty countries now have coastal-zone management plans, only a handful actually have management regimes that are in place and working. The majority of developing countries, however, have not even got to the planning stage. For instance in Kenya, of all the NGOs and community organizations formed in the country over the past several decades to combat environmental and resource degradation, not one exists which deals with coastal-resource problems and the needs of poor fishing communities.

This largely unnoticed crisis must begin to receive the same kind of attention accorded the loss of tropical rainforests and agricultural land, and the problems of desertification. Even some of UNEP's regional-seas programmes are in danger of neglect, as other, more immediate economic concerns head government agenda.

Resource management in general seems to be regarded as an issue secondary to economic development. This must change. Integrated planning of resource use and the implementation of viable coastal-zone management programmes are absolutely fundamental if we are to be able to preserve even a fraction of the rich diversity of plants and animals found in estuaries, mangrove forests, seagrasses and coral reefs. And most importantly, workable management regimes for critical coastal-zones, which involve the integration of land and sea concerns, will help provide the world's poorest billion people with sustainable livelihoods.

In order to ensure that coastal people are able to make a living from the sea, land management must merge with coastal-zone management strategies. The two must be seen as integral parts of one system. This is one of the main lessons that can be learned from UNEP's regional-seas programmes over the past 15 years.

Throughout much of the developing world, efforts at managing resources often have to contend with continually-expanding populations. The tension between human numbers and the resources needed to sustain them will become more acute in the remaining years of the twentieth century, and beyond. At current growth rates, one billion people are added to the world's population every 11 years: sometime during 1998 there will be six billion people in the world. Though fertility rates are declining slowly, the momentum of population growth ensures that at least another 2.2 billion people will be added to the planet between 1988 and 2025. The figure could be as high as 4 billion. If current trends are not reversed or at least slowed down, global population could be close to 14 billion by the year 2100.

Over 90 per cent of this growth is taking place in the developing world. Between now and the turn of the century, the population of industrialized countries will grow by only 56 million, or 5.2 per cent, while that of developing countries will grow by over 900 million, or 24.6 per cent.

It is the failure to deal with population growth, together with resource depletion, that is the underlying cause of the malaise that affects coastal areas throughout the developing South. Until the realities of population growth, uneven distribution and resource management are addressed appropriately and with vigour coastal resources will continue to deteriorate.

Although much remedial work is going on, most of it is local and small-scale. Still, UNEP's regional-seas framework offers one of the best opportunities for participating governments to forge national, as well as regional, coastal-zone management regimes. Within this framework, it is imperative that the following measures be taken during the 1990s and beyond:

1 Third World countries bordering on seas must begin to develop national strategies which integrate population policies with resource management. The two issues are inextricably linked and it makes little sense to separate them.

2 The governments which have joined together in regional-seas programmes under UNEP should devote much more attention, personnel and money to coastal-zone management.

3 Coastal-zone management must be integrated with land management if any progress is to be made in solving the coastal

effects of the following poor land-use practices: uncontrolled logging of watersheds, stripping coastal areas of vegetation for establishment of agricultural land, mining in coastal zones.

4 Within the context of coastal management, it is important that local communities and NGOs be given a strategic role in conserving resources and carrying out management programmes. Without active local support for conservation and management plans, they will never succeed.

5 Environmental education must include coastal zone problems and concerns. Without building up commitment to conservation through basic education, future generations may well repeat the mistakes of the past.

The extent to which these measures are undertaken during the next decade will determine, to a great extent, the kind of world future generations will live in. We have the chance to make a crucial difference in the quality of life for billions of people born next century. The consequences of our failure could be unimaginable.

More immediately, corrective steps must be taken now to halt population growth and coastal-resource loss so that millions of poor people will have the ability to live "sustainably", that is to say, in harmony with their environment to the extent that they do not destroy the resources future generations require. The aim of coastal management should be to give people dependent on the sea a "sustainable present". Future needs will then be easier to meet.

Bibliography

Ambio, special issue: "Marine Mammals", vol. 15, no. 3, 1986.

Ambio, special issue: "East Asian Seas", vol. 17, no. 3, 1988.

Australian Science Magazine, "The Crown of Thorns Starfish" (Great Barrier Reef Marine Park Authority, 1987), issue 3, pp. 15–51.

Baines, G., *Mangrove Resources and their Management in the South Pacific*, Topic Review no. 5 (Noumea, New Caledonia: South Pacific Regional Environment Programme, March 1981), pp. 1–7.

Berbano, M., *The Establishment of a Regime of Equal Right of Access and Non-Discrimination Regarding Transfrontier Pollution from Offshore Operations Within ASCOPE and Other Countries of the Region* (paper presented at the Seventh Meeting of the Coordinating Body on the Seas of East Asia, Yogyakarta, Indonesia, 17–19 July 1988).

Centre for Research of Human Resources and the Environment, the International Federation of Institutes for Advanced Study, *Jakarta Bay Management Study; Phase 1 – Draft Final Report* (University of Indonesia, Jakarta: CRHRE, February 1987), pp. 1–56.

Chansang, H., P. Boonyanate, M. Charuchinda, "Features of Fringing Reefs in Shallow Water Environments of Phuket Island, the Andaman Sea", in *Proceedings of the Fifth International Coral Reef Congress* (Tahiti, 1985), Vol. 6, pp. 439–44.

Clarke, W. and J. Morrison, "Land Mismanagement and the Development Imperative in Fiji" in *Land Degradation and Society* (London and New York: Methuen, 1987), pp. 176–85.

Dahl, A. L. and I.L. Baumgart, *The State of the Environment in the South Pacific*, UNEP Regional Seas Reports and Studies no. 31 (Geneva; UNEP 1983).

D'Croz, L., *Survey and Monitoring of Marine Pollution in the Bay of Panama*, unpublished paper (University of Panama, 1988).

Durning A., *Action at the Grassroots: Fighting Poverty and Environmental Decline*, Worldwatch Paper no. 88 (Washington DC: Worldwatch Institute, January 1989).

The Economist, "The Gulf: Strip Mining", 24 December 1988, pp. 71–2.

The Economist, "The Mediterranean: Cleaner, by Jove", 13 August 1988, p. 37.

The Economist, "Horribly Helpless Haiti", 14 May 1988, p. 84.

Gomez, E.D., *The Present State of Mangrove Ecosystems in Southeast Asia and*

and the Impact of Pollution – Regional (Manila: South China Sea Fisheries Development and Coordinating Programme, March 1980), pp. 1–52.

Great Barrier Reef Marine Park Authority, *AMSTAC Report on Marine Science and Technology in the Tropics: Management of the Great Barrier Reef*, unpublished paper (Great Barrier Reef Marine Park Authority, Townsville, Australia, September 1987) pp. 1–13.

Greenpeace, *Environmental Assessment of the Wider Caribbean Region for the United Nations Environment Programme's Caribbean Action Plan* (Washington DC: Greenpeace International, 1988), pp. 1–33.

Greenpeace, *Burnt Offerings 2: Greenpeace's Report on the Dumping of Philadelphia's Toxic Incinerator Ash in Haiti* (Washington DC: Greenpeace, February 1988), pp. 1–12.

Grier P., "Humanity's Castoffs: How much can the oceans take?", *Christian Science Monitor*, 11 May 1981, pp. 18–19.

Gupte, P., *The Crowded Earth: People and the Politics of Population* (New York and London: W. W. Norton and Co., 1984).

Hardoy J., and D. Satterthwaite, *Squatter Citizen: Life in the Urban Third World* (London: Earthscan Publications Ltd., 1989).

Harrison, P., *Inside the Third World: The Anatomy of Poverty* (London: Penguin Books, 1979).

Hinrichsen, D., "Homer's Wine-Dark Sea is Awash with Filth", *The Christian Science Monitor*, 29 March, 1978, p. 21.

Hinrichsen, D., *Our Common Future: A Reader's Guide* (London: Earthscan Publications Ltd., 1987), pp. 30–33.

Hinrichsen, D., *Meeting the Population Challenge*, (New York: United Nations Population Fund, 1989), pp. 5–10.

Hinrichsen, D., "Income-generation for women on Colombia's mangrove coast", *Populi*, vol. 16, no. 1, 1989, pp. 42–47.

Hinrichsen, D., (ed.), *Ambio* special issue: "The Caribbean", vol. 10, no. 6, 1981.

Hinrichsen, D., (ed.), *Ambio* special issue: "The Indian Ocean", vol. 12. no.6, 1983.

Hinrichsen, D., (ed.), *Ambio* special issue: "Population, Resources and Environment", vol. 13, no. 3, 1984, pp.142–74.

Hinrichsen, D., (ed.), *Ambio* special issue: "The South Pacific", vol. 13, no. 5–6, 1984.

Hulm, P., *A Climate of Crisis: Global Warming and the Island South Pacific* (Nairobi: UNEP, 1989), pp. 1–16.

Hulm, P., "The Regional Seas Programme: What fate for UNEP's crown jewels", *Ambio*, vol. 12, no. 1, 1983, pp. 2–13.

Indonesian National Development Planning Agency, *Action Plan for Sustainable Development of Indonesia's Marine and Coastal Resources* (Jakarta, April 1988).

International Center for Living Aquatic Resources Management, "Management Issues: Indonesia, the Philippines, West Indies", *NAGA: The ICLARM Quarterly*, April 1988, pp. 3–12.

International Center for Living Aquatic Resources Management, *Tropical*

Coastal Area Management: A Newsletter for Coastal Managers, Users and Researchers in the ASEAN Region, vol. 1, no. 1, October 1986, pp. 4–10; vol. 1, no. 2, December 1986, pp. 1–4.

International Center for Living Aquatic Resources Management, *Tropical Coastal Area Management: A Newsletter for Coastal Managers, Users and Researchers in the ASEAN Region*, vol. 2, no. 1, April 1987; vol. 2, no. 3, December 1987, pp. 2–10.

International Center for Living Aquatic Resources Management, *Tropical Coastal Area Management: A Newsletter for Coastal Managers, Users and Researchers in the ASEAN Region*, vol. 3, no. 1, April 1988.

International Center for Living Aquatic Resources Management, "Small-Scale Fisheries: conflicts and alternatives", *ICLARM Newsletter*, vol. 3, no. 3, 1980.

International Center for Living Aquatic Resources Management, "Tilapia, coral reefs, fish aging, muro-ami", *ICLARM Newsletter*, vol. 8, no. 1, January 1985.

Jacobson, J., *Environmental Refugees: A yardstick of habitability*, Worldwatch Paper 86, (Washington DC: Worldwatch Institute, November 1988).

Kamm, H., "Polluted Mediterranean on the way to recovery", *New York Times*, 21 October 1986.

Kelleher G., *Managing the Great Barrier Reef*, unpublished paper (Great Barrier Reef Marine Park Authority, Townsville, 1988).

Kelleher, G., and R. Kenchington, *Policy for Marine Conservation and Guidelines for the Establishment of Marine Protected Areas* (Great Barrier Reef Marine Park Authority, Townsville, June 1988), pp. 6–56.

Kiravanich, P., *Water Resources and Water Quality Management in Thailand*, unpublished paper (Bangkok: Ministry of Science, Technology and Energy, July 1987).

Kullenberg, G., *The Vital Seas: Questions and answers about the health of the oceans* (Geneva: UNEP, April 1984).

Kuwait Institute for Scientific Research, *Annual Report 1985–86* (Kuwait, 1986), pp. 12–29.

Lewis, N., "The happy ant heap", *The Independent Magazine*, 2 September 1989, pp. 40–44.

MacKenzie, D., "If you can't treat it, ship it", *New Scientist*, 1 April 1989, pp. 24–5.

Maragos J., A. Soegiarto, E. Gomez, M. Dow, "Development planning for tropical coastal ecosystems", in *Natural Systems for Development: what planners need to know* (Honululu: East–West Center, 1983), pp. 229–98.

Mariano, D-L, *Letter from Lingayen: The seeds of an ecological disaster*, series no. 1 (Manila: The Tambuyog Development Center, 1985), pp. 1–8.

McLarney, W.O., "Still a Dark Side to the Aquarium Trade", *International Wildlife*, March–April 1988, pp. 46–51.

McManus, L. T. and J. W. McManus, *Coral Reef Resources: Issues and recommendations*, unpublished paper submitted to the Policy Advisory Group on Conservation of Fisheries and Aquatic Resources Task Force (Manila: Philippine Department of Environment and Natural Resources, February 1987).

McManus L. T., *Coral Resources of the Philippines*, unpublished paper (University of the Philippines, Manila: Marine Science Institute, 1988).

McManus, J. W. and C. C. Arida, *Philippine Coral Reef Fisheries Management*, unpublished paper (University of the Philippines, Manila: Marine Science Institute, 1988).

Meith, N., *High and Dry: Mediterranean climate in the twenty-first century* (Nairobi: UNEP, May 1989).

Milliman, J., J. Broadus and F. Gable, "Environmental and economic implications of rising sea level and subsiding deltas: the Nile and Bengal examples", *Ambio*, vol. 18, no. 6, 1989, pp. 340–5.

Mintzer, I., "Hot Times Ahead", *People*, vol. 17, no. 1, 1990.

Mosher, L., "At Sea in the Caribbean?", in *Bordering on Trouble* (Bethesda, Maryland: Adler and Adler Publishers, 1986), pp. 235–69.

Mosher, L.,"Amazing boon: Waves of passengers set Caribbean ports jumping", *Americas*, vol. 41, no. 2, 1989, pp. 34–40.

Myers, N., *The Sinking Ark* (Oxford: Pergamon Press, 1979).

National Environmental Protection Council, *The Philippine Coastal Zone, Vol. 1: Coastal zone resources* (Quezon City: NEPC, 1980).

Newsweek, "Our Ailing Oceans", 1 August 1988, pp. 36–43.

Phantumvanit, D. and W. Liengcharernsit, "Coming to terms with Bangkok's environmental problems", *Environment and Urbanization: Environmental Problems in Third World Cities*, vol. 1, no. 1, April 1989, pp. 31–9.

Philippines National Environment Protection Council, *Philippine Environment Report 1984–85* (Manila: NEPC, December 1986). Note: the council has been turned into the Department of Environment and Natural Resources.

Porter, G., *Resources, Population, and the Philippines' Future: A Case Study*, WRI paper no. 4 (Washington DC: World Resources Institute, October 1988).

Ress, P., "The Mediterranean: surely but slowly cleaner", (Athens: UNEP press feature, July 1986).

Ress, P., "Rocking the cradle of civilization: money and mentality in the Mediterranean" (Athens: UNEP Co-ordinating Unit for the Mediterranean Action Plan, press release, June 1987).

Ress P., "Safer water, cleaner beaches and a sense of Mediterranean identity" (Geneva: UNEP press release, August 1988).

Richardson, M., "Pacific alarmed by US waste plan", *International Herald Tribune*, 14 November 1988, p. 7.

Robie D., "Rising Storm in the Pacific", *Greenpeace*, vol. 14, no. 1, January–February 1989, pp. 6–10.

Ruiz, M. A., *The Status of Mangroves in the Wider Caribbean*, unpublished paper submitted to the Intergovernmental Oceanographic Commission and the United Nations Environment Programme, October 1987.

Sadik, N., *Safeguarding the Future*, (New York: United Nations Population Fund, 1989).

Scott, D. C., "The creature that eats the Great Barrier Reef", *Christian Science Monitor*, international edition, 5–11 September 1988, p. 2.

Sestini, G., L. Jeftic and J. D. Milliman, *Implications of Expected Climate Changes in the Mediterranean Region: An Overview*, UNEP Regional Seas Reports and Studies no. 103 (Nairobi: UNEP, 1989).

Shaw, P., *Paradox of Population Growth and Environmental Degradation*, unpublished paper presented at the American Association for the Advancement of Science annual meeting, San Francisco, January 1989.

Shuaib, H. A., "Oil, development, and the environment in Kuwait", *Environment*, vol. 30, no. 6, July–August 1988, pp. 18–20 and 39–44.

Sinag-Agham, "Where Have all the Mangroves Gone?", Malaysia vol. 1, no. 2, April–June 1983, pp. 27–38.

South Pacific Regional Environment Programme (SPREP), *Country Report no. 4, Fiji*, (Noumea, New Caledonia: South Pacific Commission, August 1980).

SPREP, *Country Report no. 10, Papua New Guinea* (Noumea: South Pacific Commission, August 1980).

SPREP, *Coastal and Inland Water Quality in the South Pacific*, topic review no. 16 (Noumea; South Pacific Commission, December 1984).

SPREP, *Land Tenure and Conservation: Protected Areas in the South Pacific*, topic review no. 17, (Noumea: South Pacific Commission, January 1985).

SPREP, *Traditional Environmental Management in New Caledonia: A Review of Existing Knowledge*, topic review no. 18 (Noumea: South Pacific Commission, March 1985).

Tejada, S.M., "Staying afloat: subsistence fishery in the Philippines", *Oceans*, February 1987, pp. 46–52.

Thia-Eng, C., *Reconciliation of Coastal Resource Use Conflicts in Southeast Asia* (Manila: International Center for Living Aquatic Resources Management, April 1988), pp. 1–9.

Timberlake, L., *Only One Earth: Living for the Future* (London: BBC Books and Earthscan, 1987), pp. 73–84.

Tridech, S., *Marine Pollution Control and Management in Thailand*, unpublished paper, (Bangkok: National Environment Board, 1988).

United Nations Economic and Social Commission for Asia and the Pacific, *Training and Fellowship Requirements and Facilities in the Field of Marine Environment Protection in the ESCAP Region* (Bangkok: ESCAP February 1985).

United Nations Economic and Social Commission for Asia and the Pacific, *Environmental and Socio-Economic Aspects of Tropical Deforestation in Asia and the Pacific* (Bangkok: ESCAP 1986).

United Nations Economic and Social Commission for Asia and the Pacific, *Coastal Environmental Management Plan for the West Coast of Sri Lanka: Preliminary Survey and Interim Action Plan* (Bangkok: ESCAP, 1985).

UNEP News, "Cleaning up South Asia's seas", March–April 1987, pp. 6–7.

UNEP, *The Siren*, no. 31, July 1986, pp. 2–26.

UNEP, *The Siren*, no. 32, November 1986, pp. 11–14.

UNEP, *The Siren*, no. 33, May 1987, pp. 1–29.

UNEP, *The Siren*, no. 36, April 1988, pp. 4–36.
UNEP, *The Siren*, no. 38, October 1988, pp. 1–34.
UNEP, *The Siren*, no. 39, December 1988, pp. 1–30.
UNEP, *The Siren*, no. 41, July 1989, pp. 17–20.
UNEP, *State of the Marine Environment in the South–East Pacific Region* (Bogota: CPPS, 1988).
UNEP, *Development and Environment in the Wider Caribbean Region: A Synthesis*, UNEP Regional Seas reports and studies no. 14 (Geneva: UNEP, 1982).
UNEP, *The Health of the Oceans*, UNEP Regional Seas reports and studies no. 16 (Geneva: UNEP, 1982).
UNEP, *Pollution and the Marine Environment in the Indian Ocean*, UNEP Regional Seas reports and studies no. 13 (Geneva; UNEP, 1982).
UNEP, *Co-operation for Environmental Protection in the Pacific*, UNEP Regional Seas reports and studies no. 97 (Nairobi: UNEP, 1988).
UNEP, *The East Asian Seas Action Plan: Evaluation of its Development and Achievement*, UNEP Regional Seas reports and studies no. 86 (Nairobi: UNEP, 1987).
UNEP, *Report of the Executive Director of UNEP on the Implementation of the East Asian Seas Action Plan in 1987–88* (Nairobi: UNEP, 1988).
UNEP, *Oil Pollution and its Control in the East Asian Seas Region*, UNEP Regional Seas reports and studies no. 96 (Nairobi: UNEP, 1988).
UNEP, *State of Marine Environment in the East Asian Seas Region* (Nairobi: UNEP, 1988).
UNEP, *Coastal and Marine Environmental Problems of Somalia*, UNEP Regional Seas reports and studies no. 84 (Nairobi: UNEP, 1987).
UNEP, *Environmental Problems of the Marine and Coastal Area of Sri Lanka: National Report*, UNEP Regional Seas reports and studies no. 74 (Nairobi: UNEP, 1986).
UNEP, *Hazardous Waste Storage and Disposal in the South Pacific*, UNEP Regional Seas reports and studies no. 48 (Geneva: UNEP, 1984).
UNEP, *Management and Conservation of Renewable Marine Resources in the East Asian Seas Region*, UNEP Regional Seas reports and studies no. 65 (Geneva: UNEP, 1985).
UNEP, *The Marine and Coastal Environment of the West and Central African Region and its State of Pollution*, UNEP Regional Seas reports and studies no. 46 (Geneva: UNEP, 1984).
UNEP, *Environmental Problems of the Marine and Coastal Area of Bangladesh: National Report*, UNEP Regional Seas reports and studies no. 75 (Nairobi: UNEP, 1986).
UNEP, *Environmental Problems of the South Asian Seas Region: An Overview*, UNEP Regional Seas reports and studies no. 82 (Nairobi: UNEP, 1987).
UNEP, *Environmental Problems of the Marine and Coastal Area of India: National Report*, UNEP Regional Seas reports and studies no. 59 (Geneva: UNEP, 1985).
UNEP, *Management and Conservation of Renewable Marine Resources in the*

Indian Ocean Region: Overview, UNEP Regional Seas reports and studies no. 60 (Geneva: UNEP, 1985).

UNEP, *Pollutants from Land-Based Sources in the Mediterranean*, UNEP Regional Seas reports and studies no. 32 (Geneva: UNEP, 1984).

UNEP, *Marine and Coastal Conservation in the East African Region*, UNEP, Regional Seas reports and studies no. 39 (Geneva: UNEP, 1984).

UNEP, *Action Plan for the Protection of the Marine Environment and Coastal Areas of the South-East Pacific*, UNEP Regional Seas reports and studies no. 20 (Geneva: UNEP, 1983).

UNEP, *Action Plan for Managing the Natural Resources and Environment of the South Pacific Region*, UNEP Regional Seas reports and studies no. 29, (Geneva: UNEP, 1983).

UNEP, *Regional Seas Programme in Latin America and Wider Caribbean*, UNEP Regional Seas reports and studies no. 22 (Geneva: UNEP, 1985).

UNEP, *Environmental Management Problems in Resource Utilization and Survey of Resources in the West and Central African Region*, UNEP Regional Seas report and studies no. 37 (Geneva: UNEP, 1984).

UNEP, *Onshore Impact of Offshore Oil and Natural Gas Development in the West and Central African Region*, UNEP Regional Seas reports and studies no. 33 (Geneva: UNEP, 1984).

UNEP, *Action Plan for the Caribbean Environment Programme*, UNEP Regional Seas reports and studies no. 26 (Geneva: UNEP, 1983).

UNEP, *Conservation of Coastal and Marine Ecosystems and Living Resources of the East African Region*, UNEP Regional Seas reports and studies no. 11 (Geneva: UNEP, 1982).

UNEP, *Radioactivity in the South Pacific*, UNEP Regional Seas reports and studies no. 40 (Geneva: UNEP, 1984)

UNEP, *Marine Pollution in the East African Region*, UNEP Regional Seas reports and studies no. 8 (Geneva: UNEP, 1982).

UNEP, *Public Health Problems in the Coastal Zone of the East African Region*, UNEP Regional Seas reports and studies no. 9 (Geneva: UNEP, 1982).

UNEP, *Oil Pollution Control in the East African Region*, UNEP Regional Seas reports and studies no. 10 (Geneva: UNEP, 1982).

UNEP, *GESAMP: Cadmium, Lead and Tin in the Marine Environment*, UNEP Regional Seas reports and studies no. 56 (Geneva: UNEP, 1985).

UNEP, *Research on the Effects of Pollutants on Marine Communities and Ecosystems*, MAP Technical Reports series no. 5 (Athens: UNEP–MAP, 1986).

UNEP, *UNEP Focus*, vol. 2, no. 2, autumn 1986.

UNEP, Mediterranean Co-ordinating Unit, *Mediterranean Action Plan* (Athens: UNEP–MAP, 1985).

UNEP, *MedWaves*, no. 3 (Athens: MAP Co-ordinating Unit, 1988).

UNEP Caribbean Regional Co-ordinating Unit, *Action Plan for the Caribbean Environment Programme – A Framework for Sustainable Development* (Kingston, Jamaica: UNEP Caribbean Regional Co-ordinating Unit October 1987), pp. 7–24.

UNEP, *Implications of Climatic Changes in the South-East Pacific Region* (UNEP, CPPS, Bogota: UNEP and CPPS, April 1988).

Woods Hole Oceanographic Institution, special issue of *Oceanus*: "The Great Barrier Reef: science and management" vol. 29, no. 2, summer 1986.

World Resources Institute and International Institute for Environment and Development, *World Resources 1986* (New York: Basic Books, 1986).

World Resources Institute and International Institute for Environment and Development, *World Resources 1987* (New York: Basic Books, 1987).

World Resources Institute, International Institute for Environment and Development, and UNEP, *World Resources 1988–89* (New York: Basic Books, 1989).

Yap H. T. and E. D. Gomez, "Aspects of Benthic Recruitment on a Northern Philippine Reef", in *Proceedings of the Sixth International Coral Reef Symposium* (Manila, 1988).

Yosuke, F. "Fat prawns for Japan, slim pickings for the fisherpeople", *Japan–Asia Quarterly Review*, vol. 18, no. 4, 1986, pp. 12–22.

Index

THE AUTHORS

TOM WILBER investigates documentation regarding U.S. detainees in the Democratic Republic of Việt Nam from 1964 until 1973. His research is the source for the 2015 Hà Nội National Film Festival award-winning documentary, *The Flower Pot Story*, produced by Ngọc Dũng. A visiting lecturer at Hà Nội University in 2018, his opinion pieces have been published in *Việt Nam News*. Wilber represents a U.S.–based nongovernmental organization that works on humanitarian projects with Vietnamese organizations.

JERRY LEMBCKE grew up in Northwest Iowa. He was drafted in 1968 and served as a chaplain's assistant in Vietnam. He is the author of eight books including *The Spitting Image*, *CNN's Tailwind Tail*, and *Hanoi Jane*. His opinion pieces have appeared in *The New York Times*, *Boston Globe*, and *The Chronicle of Higher Education*. He is Associate Professor of Sociology, Emeritus, at Holy Cross College, and Distinguished Lecturer for the Organization of American Historians.

From Vietnam's Hoa Lo Prison to America Today

Dissenting POWs

Tom Wilber and Jerry Lembcke

MONTHLY REVIEW PRESS
New York

Library of Congress Cataloging-in-Publication data
available from the publisher

ISBN paper: 978-158367-908-1
ISBN cloth: 978-158367-909-8

Cover photo courtesy of Chu Chi Thanh; from Chu Chí Thành, *Memories of the War*, (Hanoi: Vietnam News Agency Publishing House, 2015).

Typeset in Minion Pro and Impact

MONTHLY REVIEW PRESS, NEW YORK
monthlyreview.org

5 4 3 2 1

Contents

Dedicated to Al Riate and Bob Chenoweth

May their persistence in conscience and inspiration to others while POWs encourage their successors in uniform to perform with comparable integrity

Former POWs Al Riate (left) and Bob Chenoweth
post-release in 1974 at Los Angeles International Airport
(photo provided by Bob Chenoweth)

Acknowledgments

Jerry is grateful for Tom Wilber's insight that the story of POW dissent needed to be written. His contribution to the book was accomplished with Tom's patience with his own impatience with IT and Carolyn Howe's help with online sourcing. Mike Yates and the Monthly Review Press team provided some best-ever editing.

Along with Jerry's appreciation for the Monthly Review team, Tom is grateful to Cora Weiss for encouragement and introductions, to Chuck Searcy for the suggestion to "talk to Jerry Lembcke," and to Jerry for listening carefully for the signal within the noise. Madame Nguyễn Thị Bích Thủy and Lê Đỗ Huy opened doors that led to unimaginable discoveries.

Introduction

On Memorial Day 2012, President Barack Obama called for the commemoration of Vietnam War events from 1961 to 1973 with these words: "Today begins the fiftieth commemoration of our war in Vietnam." The President's invitation inspired conferences, newspaper columns, and books recalling the 1965 landing of Marines at Danang and the campus teach-ins it spawned, the 1967 March on the Pentagon, and the 1968 My Lai Massacre. The documentary *The Vietnam War*, produced for public television by Ken Burns and Lynn Novick in 2017, brought interest in the war back to levels it had not had since the early postwar years.

Along with the Moratorium Days of 1969, the invasion of Cambodia and Kent State shootings of 1970, and the Christmas bombings of 1972, it is certain that interest in remembrances of the war will remain high through the fiftieth commemoration of the signing of Peace Accords in 2023, and beyond.

THE POW STORY

American prisoners of war (POWs) were made up of ground troops who were captured in South Vietnam and taken to Hanoi and pilots shot down over North Vietnam. They became critical figures in the negotiations that led to the end of the war—President Nixon insisting that U.S. troops would not be withdrawn from the South until the POWs were released, and the communist representatives insisting that there would be no prisoner return until the United States pulled out.

Concerned Americans of all political stripes rallied in support of the welfare of the POWs and in support of their families, anxious about the whereabouts of their loved ones. The Nixon administration, responding to the growing public concern for the POWs and having campaigned on a platform of ending the war, vaulted to the front of the growing parade, seizing the POW issue and weaponizing it as a negotiation lever to delay the war's end. The public responded with patriotic and humanistic concern for POW welfare. A California student group sold metal bracelets for a couple of dollars each with the POW's name, rank, and date of capture or disappearance etched on them; five million Americans bought the bracelets and vowed to wear them until the namesake returned or was accounted for. Businessman Ross Perot, to bring attention to the POW plight, bankrolled a chartered planeload of packages and mail to the captives. The Democratic Republic of Vietnam ended up refusing the shipment for logistical reasons, but Perot's effort proved to be a public relations victory, creating sympathy for the POWs and generating support for the hard line that the Nixon administration was taking in negotiations to end the war.[1]

Women Strike for Peace (WSP), meeting with the Vietnamese Women's Union in Toronto in 1969, began a process, women's group to women's group, to transport mail between POWs and their families at home in both directions, using delegations of peace activists as couriers. The WSP group returned just before Christmas 1969 with 138 letters from 132 prisoners and a provisional list of Americans held in Hanoi. Under the name Committee of Liaison with Families of Servicemen Detained in North Vietnam, the service delivered thousands of letters by the end of the war.[2]

After a gradual escalation in the late 1950s through 1964, the American war in Vietnam lasted nearly a decade, with massive increases beginning in 1965. The standard tour of duty for military personnel was twelve months, which meant there was a constant churn of those returning home and their newly deployed replacements. The country's emotions were divided: pride and elation for the returnees, grief for those who would never come home, worry for the wounded, and fear for those leaving for the war zone. With the comings and goings strung out for ages, emotions were uncentered when the Peace Accords were signed in January 1973. Many of the veterans had been home for years by the war's end, and troop levels had dwindled to less than 10 percent of the 1968 peak. These factors, when coupled with the war's loss, meant that welcome-home parades and their accoutrements were not in the offing—at least not until the POWs were released and returned in February and March of 1973.

Five hundred and ninety-one prisoners of war were released in the weeks following the Peace Accords that ended the war. The obscurity of the POWs' imprisonment and even questions about the survival of some of them lent an air of enchantment to the figures stepping off the planes at Clark Air Force Base. With bated breath, Americans of all persuasions anticipated the stories of the POWs' experience that would confirm the villainy of their captors and the strength of their own determination to return with honor, as affirmation of the goals seeded and nurtured by the Nixon administration since early 1969.

Dubbed "Operation Homecoming" by the Pentagon, the reception given the POWs surpassed anything mounted for previous generations of war veterans. Tickertape parades in New York City and Dallas produced the iconic scenes for returning warriors usually made in Hollywood; horn-honking caravans and marching bands typically accorded state-championship high school sports teams ushered POWs into small towns as conquering heroes; newspaper headlines kept the POWs in the spotlight for weeks. In May 1973, President Nixon welcomed them and their spouses to the White House in the largest reception of its type to date.

Theirs were cover-page portraits for all the major news magazines, their survival an allegory for the military victory otherwise denied, their return igniting triumphalist energies capped by years of disappointing news from the war front.

THE OTHER VOICES COMING HOME FROM HANOI

But amid the celebrations, there was discord. A March 16, 1973, *New York Times* headline announced, "Eight May Face Courts-Martial for Antiwar Roles as P.O.W.s." The eight enlisted men were accused of having expressed their opposition to the war while held as prisoners. Days later, one of two officers who would also face charges, Navy captain Gene Wilber (author Tom Wilber's father), shot down over North Vietnam in 1968, was grilled by journalist Mike Wallace for CBS's *60 Minutes* about the antiwar statements he had made. Had Wilber succumbed to torture? Or had his dissent been bought by prison guards for favorable treatment? Wilber parried Wallace's inferences and ended the interview saying his words had been sincere expressions of his "morality and conscience."

NEW SOURCES ON DISSENT WITHIN HOA LO PRISON

This book tells the untold story of POW dissent and develops the history of attempts to repress that dissent and purge it from public memory. Tom Wilber's interviews with more than a dozen former administrators, guards, and staff workers from the notorious "Hanoi Hilton" supplement primary documents he found on more than thirty investigative trips to Vietnam that support a challenge to conventional historical accounts of the POW experience and debunk the legends that have grown around it.

Chief among Tom Wilber's findings are POWs' handwritten reflections on their roles and the U.S. government's role in the conflict, letters from family members, texts for recording sessions by POWs broadcast on Radio Hanoi, some of which have been correlated to CIA-monitored transcripts of broadcasts matching those same documents, handwritten English-language training books and materials authored by prisoners, dozens of editorial cartoons by prisoners, and a range of international reading materials from the Hoa Lo prison library. On the basis of individual interviews of former detention camp staff, Wilber describes internal camp processes, including how statements were taken from prisoners, documented, recorded, and broadcast. From this research, he learned even the mundane routines of prison life, such as budgets, supply logistics, food procurement and preparation, and camp regulations.

CLASS: THE ROOTS OF DISSENT

The struggle between antiwar POWs and their Senior Ranking Officers involved the dissidents' conscientious objections to the war and their resistance to the controlling behavior of the SROs. The SROs saw their exercise of authority as legally vested in their rank and supported by their reading of the Code of Conduct. Threatening POW dissenters with courts-martial for their acts after their release, the SROs were able to intimidate them enough to ply the news media with their version of imprisonment, a version that became known as the "official story."

Left that way in the existing literature, the story of conflict among the POWs is understandable with conventional models of organizational behavior—for example, the labor-management relations common to most workplaces—tempered with the range of political and social values specific to war and military service.

In *Dissenting POWs*, however, we pursue a hunch that the tensions between POWs were rooted in the disparate socioeconomic backgrounds of the antagonists. The privileged backgrounds of the SROs were in sharp contrast with the modest origins of war resisters. It's a hunch triggered by clues scattered by Craig Howes in his 1993 book *Voices of the Vietnam POWs,* and reinforced in Milton Bates's 1996 *The Wars We Took to Vietnam.* We use newly available biographical and oral history material to show that class disparities extended into the SRO ranks. Objections to the war voiced by two of the most senior officers, Gene Wilber and Edison Miller, got them banished by their peers.

CLASS: A DISCOURSE FOR STIGMATIZING AND DISPLACING DISSENT

Just as "class" designates an objective social position with implications for wealth and income and values derived from its material realities, it also connotes the subjective evaluations that members of one class make of others. Modifiers like "upper" and "lower" class imply character and even moral judgments that are then arranged, even if unwittingly, into hierarchies for assigning social standing and status.

The first substantial history of the POW experience, John Hubbell's *P.O.W.: A Definitive History of the American Prisoner-of-War Experience in Vietnam, 1964–1973*, records that the SROs viewed the antiwar expressions of their POWs within a class framework: the rebels were from poor or broken families, less educated than themselves, and of weaker personal character. The "weak character" rap against the antiwar captives was a way to dismiss the authenticity of their political views, a form of "psychologizing the political" that *Dissenting POWs* will show was integral as well to the way that GI and veteran dissent was being categorized stateside.

The mental health discourse deployed by the SROs against their comrades in confinement had been prototyped in Albert Biderman's 1963 *March to Calumny: The Story of American POWs in the Korean War* and would come to dominate the narrative of POWs returned from Hanoi in early 1973.

SILENCING DISSENT: POWERS OF STATE, MEDIA, AND NARRATIVE

Even the best histories of U.S. prisoners of war in Vietnam, books by Michael J. Allen, Elliot Gruner, H. Bruce Franklin, and Natasha Zaretsky, give slight attention to the story of dissidents within the POW population, and even less attention to why public memory of those voices is lost. In *Dissenting POWs* we restore to proper prominence the record of antiwar voices within the POW population. We cull from the existing literature the role of government and Pentagon censorship in suppressing the story, and the role of Hollywood in eliding altogether the presence of antiwar POWs from its scripts and otherwise milking their story for its entertainment value à *la* the Rambo series and POW-rescue films.

However, we argue that the public memory of dissenting POWs was lost less to censoring or the demands of the movie market than it was to displacement of their acts of principled courage by images of them as victims of the war. Casting POW dissenters as sadsack losers precluded their inclusion in the great American captivity narrative at the center of the nation's founding mythology, wherein the captive-hero remains a prisoner *at* war, loyal to the mission on which he was sent. The antiwar Vietnam POWs would be the antiheroes in, say, the legend of John Smith and his resolve in the face of torture and the temptations of Pocahontas—the weakling POWs had no place in the American story, no association with traditions out of which memories could have been constructed.

The *coup de grâce* to American memory of the POW experience was dealt by the rumors that some POWs had been left behind when the United States departed Vietnam. The conspiratorial threads in those stories had it that prisoners had been knowingly abandoned by shadowy sell-out forces within the U.S. government that settled the war on terms favorable to international communism. *Dissenting POWs* delves into the post–Vietnam War POW-MIA fantasies that echoed the Cold War hysteria about brainwashing and prisoner defections.

The banishment of POW dissent from memory leaves a void in American political culture where new generations of uniformed war resisters will look for role models, and civilian activists will look for allies in their efforts to end U.S. wars of aggression. Filling that void is the mission of this book.

1

Forgotten Voices from Hoa Lo Prison: Dissent in the Hero-Prisoner Story

At first glance, the history of U.S. POWs in Vietnam appears to be an exercise in hero construction, a story of prisoners remaining loyal to their mission and one another. Many of their memoirs record the resistance of shot-down pilots to efforts at extracting sensitive information from them in return for better medical care, food, and early release.[1] When they defied their captors, some of the POWs say they were tortured until they complied. The hero-prisoner was a holdout who sacrificed his own comfort rather than divulging information that might put his comrades, in the air or field, at greater risk or even compromise the security of the American homeland.[2]

The hardcore holdout-POW is the central figure in the history of the American version of the Vietnam War POW experience that historian Craig Howes sarcastically dubbed the "Official Story."[3] The official story appeared in print in 1976 as *P.O.W.: A Definitive History of the American Prisoner of War Experience in Vietnam, 1964–1973.* The book was a project of *Reader's Digest,* proposed to author John G. Hubbell by the magazine's managing editor, Kenneth O. Gilmore, who gave him a "blank check" for the work.[4] The Pentagon assigned its public affairs chief, Jerry Friedheim, to assist Hubbell. Friedheim also had key roles in managing the Operation Homecoming messaging and granting access to former POWs for interviews and press releases, as well as managing information requests from congressional staffs related to POWs changing their stories from fair treatment to torture after their release. Hubbell's *P.O.W.* also had the imprimatur of Admiral Thomas H. Moorer, the Chairman of the Joint Chiefs of Staff (JCS), virtually ensuring that *P.O.W.* would reflect the interests of the military establishment. Nobody was more establishment than Moorer and nobody had a greater stake in the way the POW story would be told. He was Chief of Naval Operations from 1967 to 1970, during which time, in '68 and '69, 116 Americans were captured, many of them Navy pilots.

Moorer became Chairman of the Joint Chiefs on July 1, 1970, and assumed control of one of the most controversial and secretive operations of the entire war—the POW rescue raid on Son Tay in North Vietnam where it was thought as many as seventy U.S. prisoners were held. Son Tay, about 23 miles west of Hanoi, was one of several POW holding centers in the Hanoi region. The Vietnamese designated each of them by number and Son Tay was T142. The American POWs coined nicknames for many of the sites, such as Briarpatch, Zoo, and Skid Row. Only the central prison, Hoa Lo, was known as the "Hanoi Hilton," though it, too, was sectioned off with POW nicknames such as New Guy Village, Camp Unity, Little Vegas, and Heartbreak.

Son Tay had been under aerial surveillance for weeks when Moorer took over as JCS chairman, and planning for the raid was already underway. The plan called for about fifty Special Forces troops to launch by multiple helicopters from Thailand, led by sophisticated slow-flying navigation aircraft, to fly over Laos and land inside the prison compound. The operation involved in-flight refueling of the helicopters and the support of fighter-bombers providing cover for the raiders. A sea-launched bombing raid on Hanoi would divert attention from the attack on Son Tay, twenty-three miles west of the city. Training for the mission, costing millions of dollars, had taken place over several months at Eglin Air Force Base in Florida. The raiding party was made up of some of the war's most decorated soldiers, including the legendary Green Berets Bull Simons and Dick Meadows.

The raiders set off on the night of November 20, 1970, but the mission was a failure with a comedic flare. Intelligence failures were evident from the outset, including misjudging the density and the height of trees at Son Tay, which differed from the trees at their practice site. One chopper crashed after hitting a tree; Bull Simons's team of twenty-two men landed at a secondary school a quarter-mile south of the target; and, if that was not enough, Son Tay was empty of prisoners—there were no POWs there.[5] Hubbell began work for his book hand-in-hand with Pentagon insiders at the very time Moorer was enmeshed in the scandal over who was responsible for the Son Tay debacle and the conspiratorial speculations that led to it. The book's Acknowledgments say Hubbell asked for Moorer's "support and *guidance*" (emphasis added) in the writing and a few sentences later repeated the words saying Moorer's support and guidance were "forthcoming in unstinting measure." The suggestion in those words of Hubbell's subservience to the Pentagon's need for a whitewashed POW history gains confirmation in the book's later assessment of the Son Tay raid as "a huge success."

> *I like people who weren't captured.*
>
> —Donald Trump

Donald Trump is responding here to Arizona senator John McCain, who had earlier expressed his own skepticism of Trump's suitability for the presidency. The campaigns for the 2016 elections were just getting underway, and McCain supporters had been invoking the senator's record as a fighter pilot and POW during the war in Vietnam. McCain's five and a half years as a prisoner in the legendary Hanoi Hilton, they said, made him an American hero. Trump's retort made headlines on July 18, 2015, because of its use for campaign leverage, but it was a line he had been using for years.[6]

The centrality of the POW story to American remembrance of the war in Vietnam and the competing narratives it helps construct could not be made clearer than does this give-and-take between candidate Trump and Senator McCain's supporters. The image of pilots shot out of the sky, captured and rendered helpless, and ground troops taken prisoner and held in bamboo cages, were symbolic of the lost war itself. Their confinement symbolized the inferiority of the U.S. military men and machines, and the humiliation they faced on the international stage. It was exactly the imagery that Moorer and the *Reader's Digest* team had in mind, and which they wanted expunged from memory, when they commissioned John Hubbell to create the alternative "official" story out of the actual facts of shot-down pilots and captive GIs.

The hero-prisoner story would portray the prison experience as, in a sense, another war front, another theater-of-war, in which the POWs fought gallantly "behind enemy lines" and came out on top as hero-warriors. Carrier air wing commander James Stockdale (promoted after captivity to admiral rank) who was shot down on September 9, 1965, fit this profile. As a carrier air group commander who chose to fly a mission on that fateful day, Stockdale was the highest-ranking officer captured. Described by Craig Howes[7] as "one of the Navy's 'princes of the realm, the blood Royal,'" he had graduated from the Naval Academy and gotten a master's degree in international relations at Stanford University. At Stanford, he had learned how U.S. POWs had been brainwashed during the Korean War. Now, he made the association in his mind: the Vietnamese guards' compassionate treatment of captives while "educating" them on the history of the Vietnamese independence movement was an attempt at mind control. As the senior officer among the POW population, his enforcement of noncompliance among his more junior POWs had to be absolute.[8]

Hubbell did return to the brainwashing theme later in the book when describing the hard line that SROs (Senior Ranking Officers) like Stockdale were taking against their own, lower-ranking, fellow captives. However, he didn't belabor it. He may have assumed that mere references to brainwashing were sufficient given the historical closeness, at the time, of Americans to their Korean War experience. Moreover, the 1962 film *The Manchurian Candidate* starring Frank Sinatra, Janet Leigh, and Angela Lansbury had lodged into American imaginations the image of a brainwashed POW returning from Korea as a communist sleeper agent.[9]

Stockdale had command-level company in Navy Commander (later Rear Admiral) Jeremiah Denton Jr. and Air Force Lieutenant Colonel (later Brigadier General) Robinson Risner. Denton was a squadron executive officer shot down on July 18 and like Stockdale was a product of the Naval Academy with a master's degree, his in international relations from Georgetown. In his memoir *When Hell Was in Session*, he writes of his parents' pride in their "Southern aristocratic background" and touts his own expertise in airborne electronics, antisubmarine warfare, and air defense. He assessed himself a "good catch" for the North Vietnamese. Risner had been shot down just days after Stockdale and entered the prison system as a bona fide American hero: a Korean War ace and test pilot, he had set a transatlantic speed record flying the anniversary Lindbergh flight in 1957 and had been rescued after being shot down over North Vietnam months earlier. However, unlike the service academy "royals" Stockdale and Denton, Risner, son of a sharecropper, entered the Army during the Second World War after high school and trained for two years to become a pilot, at which point he was commissioned a second lieutenant and served out the remainder of the war in Panama, leaving the military after the war, while maintaining his flying skills with the Oklahoma Air National Guard. He was recalled to active duty in 1951 after hostilities broke out in Korea.

Fortunately for Stockdale, Denton, and Risner, whom Craig Howes would later dub, sardonically, "the triumvirate of great leaders," they were handed Navy Lieutenant Junior Grade Rodney Knutson, a radar intercept officer, shot down with pilot Ralph Gaither on October 17, 1965. The peculiar circumstances of Knutson's capture were fitting for the three to begin confecting a chain of command within the prison population that ran parallel to that of the North Vietnamese prison administration. Later acknowledging its "Mickey Mouse" appearance, they imagined the hierarchy they desired being justified if not required by the U.S. Military Code of Conduct for POWs.

In retrospect, the made-up chain of command may have been a function of the SROs' authoritarian personalities—their psychological need to subordinate others—or as a ruse to manage the theatrics of the prisoner-*at*-war narrative they wanted performed. By the time of Knutson's arrival, Stockdale, Denton, and Risner had already abandoned their tough-guy holdout postures and made compromising statements to their Vietnamese interrogators. They said they had been tortured into talking. But had they? The question would never be asked if all the later-arriving captives, beginning with Knutson, could tell the same story. Thusly reasoned, notification of the chain of command was disseminated through the prison along with the order from the top that *all* prisoners should be like the Great Three and submit to torture before making statements beyond the name, rank, and serial number protocol. In the end, pressures to meet the expectations of the SROs led to behavior that *invited* torture to avoid post-release charges for collaboration upon release.

RODNEY KNUTSON: A PRISONER *AT* WAR

Hubbell found the model for his still-in-the-fight captive in Navy Lieutenant Junior Grade Rodney Knutson. In Hubbell's account, Knutson was the first shot-down pilot to have exchanged gunfire with his Vietnamese captors. Hubbell wrote that Knutson had fired a tracer round from his .38 caliber pistol into the head of a Vietnamese militiaman, and a round at point-blank range at another before blacking out. Then taken to Hanoi by "a small army contingent," he was assaulted along the way by villagers who were egged on by an officer with a bullhorn.[10] At Hoa Lo, he then refused to cooperate in the questioning by interrogators beyond the standard practice of giving his name, rank, and serial number—Knutson was a prisoner, but not out of action.

Knutson's description in *P.O.W.* of his conditions of incarceration at Hoa Lo was unpleasant to put it mildly: distasteful food, endless interrogations, and rats. His description of the rats, some as large as small dogs that crawled close to his face, is graphic and chilling. But his description of the physical abuse he is meted takes readers into new territory: lying on his stomach, a length of clothesline rope was looped around his arms just above the elbow then pulled and cinched until circulation was cut off. With his elbows tied together, he was made to sit with his ankles locked into stocks at the foot of the bed. He was then ordered to apologize for "insulting the Vietnamese," and when he didn't, he was then beaten until his nose and several teeth were broken.

The brutalizing of Knutson continues for five pages of *P.O.W.*, completing, thereby, the book's transition to the first of the torture allegations that is the centerpiece of the prisoner-at-war narrative. Knutson was the first of several POWs "going to the ropes," as they phrased it.[11]

But was it torture, or punishment? The answer is a matter of the guards' motivation or intent. By Hubbell's telling, Knutson had refused to eat his dinner and dumped it into a waste bucket and then refused to apologize. Hubbell quotes Knutson as being told: "For insulting the Vietnamese people . . . you must be punished."[12] Rule 10 of "Camp Regulations" posted in each cell was: "Violations of the regulations shall be punished." Was the wasteful act of throwing away a meal a prison discipline problem?

It is curious, however, that POW refusals to eat had been routine since the first pilot taken captive. Lieutenant Junior Grade Everett Álvarez had arrived fifteen months earlier. Alvarez had flown off the carrier USS *Constellation* on August 5, 1964, in an A4 Skyhawk over the Tonkin Gulf along the coast of Vietnam. After strafing a North Vietnamese patrol boat, his plane was hit and he parachuted into the waters of Ha Long Bay some sixty kilometers east-northeast of Haiphong. Álvarez was plucked from the water by Vietnamese in a small boat and wrapped in rope "like a top" wrote Hubbell. Some of his captors screamed at him and kicked him. "Certain that he was about to be hanged by his ankles, skinned, relieved of his testicles and finally his head," he was transferred to a torpedo boat and taken to shore. On August 11, he was transported by jeep to Hoa Lo prison in Hanoi.[13]

As the first of a population that would eventually grow, although slowly at first, to nearly 600, Álvarez was likely unaware that his captors had no plan what to do with him. With a captured pilot in transport to Hanoi, the army went to the Hanoi police to ask for space in Hoa Lo Prison, at that time a civil jail administered by the city. The police gave the army some empty offices initially to use as prison cells, and then "four or five" actual cells in one section of the prison. Eventually the police transferred control of Hoa Loa to the army, but until kitchen facilities were available, the army guards went to local restaurants to buy carry-out for the initial prisoners.[14]

Still, as described in *P.O.W.*, Álvarez's prison conditions were unpleasant. Some of the meals—chicken heads, a cow hoof, animal hair, and shrimp with eyes in—he left uneaten; others he vomited up. He fed interrogators only misinformation, and yet they only played on his fears of never going home—there were no ropes for him.

So why would the Knutson incident have called forth more severe measures? Why now? Why him? Was his punishment really for insulting the Vietnamese by wasting their food and then refusing to apologize? What about the two Vietnamese militia people he had shot and maybe killed? It makes sense that someone be "sent to the ropes" as *punishment* for a crime like murder, as the Vietnamese may have viewed the shootings. But "ropes" for rudeness seems like an excuse for meanness—torture—which is the essential ingredient in the prisoner-at-war story that Hubbell is writing. Might it be the case, then, that Hubbell switched-out the criminal indictment for one of insensitivity and defiance? Hubbell's hero-prisoner story is riddled with incongruities like these.

It's probable as well that the torture spin on Knutson's story was already part of the camp lore that elevated into a canon in the hero-prisoner myth/legend that would be brought back from Hanoi by the SROs in 1973, rather than something that Hubbell created. Knutson's obstinance and toughness were, after all, exactly the traits that Stockdale could point to in October 1965 as the standard that all the captives should meet—Knutson embodied the prisoner-at-war ethic. When the two met on October 29, Stockdale supposedly said to Knutson, "I think you did a fine job, Rob . . . you took the right approach. Give them nothing, make them take it from you."[15]

Stockdale's adoption of Knutson as his prisoner-at-war poster-boy was not without complications. It was against international law for Knutson to have shot a tracer round at an enemy. Years later, he acknowledged thinking, at the moment of shooting, that he could get in trouble for it.[16] Stockdale, as his prison SRO, either didn't know about the shooting—it's possible that Knutson didn't tell him—or was complicit in reconfiguring the punishment-for-crime story (or discipline-for-breaking-camp-rules-by-wasting-food story) into the torture-for-resistance story more useful for the prisoner-at-war role he had in mind for Knutson.[17]

In covering the eighteen months after the Knutson incident, *P.O. W.* is a compilation of shoot downs, backgrounds of the pilots, the stories of their capture and transport to Hanoi, descriptions of the appalling conditions of their confinement, their despair at never going home, the endless interrogations, and the torture they suffered for refusing to answer the questions and refusing to apologize for not answering.

WRINKLES IN THE HERO-PRISONER STORY

The united front of the prisoners-at-war in Hoa Lo faced its first challenge in the spring of 1967 when news of the stateside antiwar movement reached their eyes and ears. The April 15 demonstrations across the country had brought together labor, religious, SDS (Students for a Democratic Society), and civil rights groups to oppose the war. Known as Spring Mobe, short for the Spring Mobilization to End the War in Vietnam, the effort turned out 400,000 protesters in New York City and 75,000 in San Francisco. The success of the Mobe also spurred the organizing of resistance within the military. Bus terminals like the New York's Port Authority, through which hundreds of GIs passed daily, became centers of protester outreach to military personnel. Some military members, like Private Howard Petrick stationed at Fort Hood in Texas, took the radical literature of the civilian activists into their barracks for distribution to their buddies. A year later, those efforts blossomed into a network of underground antiwar newspapers written and printed by and for GIs, Marines, sailors, and airmen.

Hoa Lo administrators made sure that news of the growing homefront opposition to the war reached the prison inmates.[18] Hubbell wrote that prison staff had POWs read newspaper reports of the demonstrations to their fellow captives over the intercom. They all heard the news that political and religious leaders were condemning the war: Martin Luther King Jr. calling it "blasphemy"; Dr. Benjamin Spock expressing "scorn and horror" for it; Nobel Prize–winner Dr. Linus Pauling feeling "shame" for the war. Hubbell wrote that POWs "felt bewildered, depressed, betrayed. They understood political leaders dissenting from policy and opposing it, but not openly opposing it to the enemy's benefit while the country was still at war."[19]

Although Hubbell presents it as such, news of the Spring Mobe would not have been shocking to the POWs. The prison administration regularly made U.S. news magazines like *Time* and *Newsweek* available to the prisoners, and those publications were full of antiwar news even before the 1967 Mobe. In November 1965, Norman Morrison, a Quaker opposed to the war for religious reasons, immolated himself outside the Pentagon office of Defense secretary Robert McNamara. On March 26, 1966 (Sunday March 27, 1966), the *New York Times* ran a striking front-page visual pairing a photograph of Marines going ashore in Vietnam with a photograph of the antiwar march. The caption atop the two photos read: "Marines Land South of Saigon—Marchers Protest Policy on Vietnam." The Bertrand Russell Tribunal had convened in Stockholm in December 1966 to consider allegations of war crimes committed by the United States. Historians Stuart Rochester and Frederick Kiley say the POWs knew about the Morrison immolation and the Russell Tribunal.[20]

Additionally, pilots shot down after March 1965 when teach-ins against the war had begun on college campuses would themselves have been exposed to the growing unpopularity of the war; many of them, after all, would have been students on those campuses. In October 1967, a delegation of peace activists from the United States met with prisoners in Hanoi. One of them, Air Force captain Larry Carrigan, was asked about his knowledge of demonstrations in the States:

> Sure, we knew; at Flight School there were two guys from Berkeley who told us about demonstrations. We figured it was something to do on a Saturday afternoon, get together and paint a sign. I remember Econ class, the guy used to sit next to me, all of a sudden next week he wasn't there, got drafted, it got you thinking.[21]

Not surprisingly, then, the later shoot downs arrived with "long hair, sideburns, beards and mustaches," styles that may have suggested an affinity with the stateside antiwar and counterculture movements that was unnerving when seen by those already in Hoa Lo.[22]

Indeed, military personnel and veterans were becoming prominent in the antiwar movement by late 1965. A November 24, 1965, advertisement in the *New York Times* sponsored by the Ad Hoc Committee of Veterans for Peace in Vietnam had excoriated the November 1965 fight for the Ia Drang Valley in which 237 Americans and 1,200 North Vietnamese had been killed. In July 1966, the refusal of three GIs at Fort Hood to report for shipment to Vietnam made news across the country.

Finally, direct contact with American peace activists visiting Hoa Lo would have alerted POWs to the escalating opposition to the war at home. In August 1965 at the World Peace Congress in Helsinki, W. E. B. Du Bois Club members, student and civil rights activist Harold Supriano, Du Bois Club international secretary Michael Myerson, WBAI-FM program director Jon Christopher Koch, and freelance writer Richard Ward were invited to come to Hanoi in the fall at the invitation of the Vietnamese delegates.[23] In the fall of 1965 these members of the American antiwar W. E. B. Du Bois club visited Hoa Lo and met with American POW Air Force captain Robert Daughtrey[24] from Texas who had been captured in August. In

December 1965, Professors Herbert Aptheker and Staughton Lynd, along with SDS president Tom Hayden, were invited to Hanoi by the North Vietnamese through the World Peace Congress. Aptheker said they "interviewed" an Air Force POW of five months, who had been shot down "on (his) first mission,"[25] and returned to the States with letters he had written to his family.[26]

The POWs who met with the peace travelers may have been the forerunners of a larger group that would emerge later to speak out against the war and resist the authority of the SROs. And there may have been more than two peace-leaning prisoners at the time. Hubbell, after all, makes no mention of Daughtrey's meeting with the Du Bois delegation and does not mention any meeting of POWs with the Aptheker group. Might he have left out other early instances of POWs conversing with antiwar activists?

Hubbell doesn't tell of the early meetings of antiwar visitors with POWs, but when he does tell of those events, he suggests that the prisoners were forced to do it by the guards.[27] His explanation that it was coercion, rather than prisoners' own belief, may have been correct. But many more prisoner-activist meetings would occur in the coming years, and historians writing in the postwar years were skeptical of the idea that they were all coerced.[28] The stories, for example, that POWs were forced to meet actor and activist Jane Fonda when she visited Hanoi in 1972 have been discredited as attempts to discount the sincerity of the POWs and vilify her and the Vietnamese. On the other hand, when coupled with Hubbell's omission of previous instances of prisoner meetups with peace activists, such as the Du Bois group, it is also possible that he was manufacturing a case, inferentially at least, that dissent was not indigenous to the prison population itself and appeared only later when media reports, biased against the war, began leaking into camp.

DISSENT BREAKS OUT IN HOA LO

Hubbell writes that by 1971 "at least 30 percent and perhaps as many as 50 percent of the prisoners were disillusioned about the war and becoming increasingly cynical about it."[29] Left at that, the numbers seem to contradict the central narrative of the hero-prisoner myth/legend, the "official story" that the POWs remained *at* war until they returned home with honor at the bitter end. For Hubbell, however, the numbers provide confirmation of two contaminating externalities at work in the culture of the prison population: the pervasive use of torture as a kind of conversion therapy by the prison guards, and the arrival in Hanoi in 1971 of U.S Army and Marine Corps ground troops who had been taken prisoner in the South and whose mental and physical capabilities made them more susceptible to communist propaganda and mind control techniques.

"LIKE 95% OF THE POWs, I WAS TORTURED MANY TIMES."[30]

The prisoner *at* war is the central figure in the hero-prisoner story, and the experience of *torture* provides the validation that what went on in the Hanoi prison system was a form of war. The credibility of the narrative hinged on the verity of the torture claims. Postwar studies raised serious questions about the claims, and the interpolation of those questions, in turn, sheds light on the origins of dissent within Hoa Lo.

One question mark on the pervasiveness of torture has long been the absence of a comprehensive account of the POW experiences. The Vietnamese always denied using torture, and despite the political realignments, defections, and emigration that have characterized Vietnam in the postwar years, no former guards or prison administrators have corroborated the charges of the former POWs. Instead of corroboration, we have a London *Times* story of October 25, 2008, for which the reporter sought out Tran Trong Duyet, the former prison director, to ask about the claim by 2008 presidential candidate John McCain that he had been tortured as a prisoner in Hanoi. Duyet said, "I never tortured or mistreated the POWs nor did my staff." Nguyen Tien Tran, another director, confirmed Duyet, saying, "We had a clear code of taking care of the injured. Why would [McCain] say he was tortured?" The prison guards do not deny physical discipline, and they point to rules violations or responses to violent outbursts of anger or frustrations by the prisoners as reasons to administer physical discipline. When asked specifically why then would former prisoners report that they were tortured, even contradicting their statements otherwise, Nguyen Minh Y, retired camp administrator, replied that "people are opportunists."[31]

The former prison director's disavowal of torture certainly can be dismissed as self-serving, but in the absence of documentary evidence that torture *did* happen, the only "record" of it is the memories of the prisoners themselves, a record marked by their own self-interest in maintaining the realism of torture in the hero-prisoner-at-war narrative. But the remembered version has too much variation to be reliable. Almost all accounts agree that there was no torture after 1969, which means that any accounts of the 156 flyers shot down after January 1, 1970 (about a quarter of the POWs that returned in February and March 1973) is marginal to the prisoner-at-war story. There is a similar marker on the early years. The twenty-three pilots shot down during the first year of the air war over the North reported that they were not tortured until September 1965, after which some say they were and some say they were not, and most have remained silent on the issue. That discrepant pattern continues through the fall of 1969, raising questions about the reliability of the reports and the reasons for the variation of the memories.

Claims of torture are also countered by civilian visitors to the prison. Carol McEldowney, an American, was there in 1967 at the supposedly peak period of torture. She wrote in her journal that the "decent treatment" of prisoners had been verified and was no longer an issue. The objection that McEldowney was an advocate for ending the war and therefore biased against U.S. policy is itself countered by other writing in her journal. She wrestled with her own bias. Calling some of the North Vietnamese presentations on non-POW matters "bullshit" and "propaganda," she was, if anything, predisposed to believe the opposite of what her observations were telling her about the POWs' conditions.[32]

There is also disparity in the memories of POWs from different branches of service: of the seventy-five Army POWs who came home in 1973, all captured before 1970, forty tell their stories in the collection *We Came Home*, edited by Barbara Powers Wyatt, but none describes having been tortured. This is not to say that some of them were not treated badly. Gustav Mehrer, for example, described having been bound and hung by his arms during interrogation, but his report was an exception and he did not refer to it as torture. Army Major William Hardy made a point of saying he had *not* been tortured. George Smith's story is not in the Wyatt volume, but he wrote in his own book that he had been treated well by the National Liberation Front (NLF), known in the United States as the Viet Cong, literally "Vietnamese Communist," in the South. At the very least, this variegated picture dispels the claim made by Hubbell in his *P.O.W.* that torture was a systematic, day-to-day policy of the Vietnamese.

As an example of the inconsistencies in the treatment narratives, David Wesley Hoffman's story is a case study. Hoffman was interviewed by George Wald on February 19, 1972, less than two months after his capture. Wald was a Harvard professor, peace activist, and Nobel Prize winner for medical research. Hoffman was effusive in the details of his medical care for the compound fracture of his arm from his ejection. It was an obvious topic for discussion as Hoffman's immobilized arm projected horizontally from a cast that hospital doctors had specially configured to correctly heal his compound fracture of the humerus. Wald asked about his treatment when captured by local people:

GW: Did they treat you all right?

DH: They took me into their village and immediately got medical treatment. It was obvious to them I was hurt and they went and got the local doctor to come. It was a woman as a matter of fact. She strapped my arm, put a temporary splint on the thing and fixed it so that it wouldn't move so that when I was transported in, it was immobile, so that I didn't have problems there. They fed me, gave me warm food, a good place for shelter, and treated me very well.

GW: Doesn't it surprise you that they treated you so well?

DH: I was amazed, frankly and honestly. I didn't know what to expect, honestly. But I was amazed, and I have been constantly and continually amazed at the treatment. From the time I was shot down until this very moment my treatment has been superb.

After some back and forth discussion why the United States should end the war and Hoffman's hope that Americans will elect a president who can end it, the conversation went back to his arm:

DH: I'm not sure whether my arm hit part of the aircraft when we ejected out of it, or whether we were going so fast that the wind blast caused the injury. I had no other injury, there was just the fracture of the left arm.

GW: It's the upper bone?

DH: Right up in here, yes sir, right up in the middle. The doctors seem pretty satisfied that when they do take the cast off, that I'll have full use of the arm and that everything will be all right again. From what I've seen of the medical treatment—and that's another thing—medical treatment. One of the doctors comes every other day and checks on us to make sure we're perfectly all right. For instance, I was having a problem with my hand when I go to sleep, the hand would droop and go to sleep. I'd wake up in the morning with it stiff. So today they brought me this (showing a little, wrapped paddle that slipped into the end of his splint) and it holds the hand up; and now when I go to sleep the hand stays straight and doesn't cut the blood off. They are constantly concerned and checking on us. They feed us more than we can eat.

Hoffman expounded more to Wald on the "very good" food and eating "better than the guards who guard us."

Were these talking points required by his captors, it is hard to imagine a better acting performance: Hoffman's extemporaneous offering of information to Wald's unrehearsed questions seem to reinforce his belief in his own words.

With similar conviction, Hoffman joined eight other captured pilots in signing an April 1972 statement to the U.S. Congress objecting to a recent escalation of bombing. Asked during a May video interview to reiterate the substance of that statement, he said the bombings were a danger to the POWs that would also likely lengthen their imprisonment and were an ineffective means to dampen Vietnamese spirits. He made an eloquent call for peace and return to peace talks and, with other prisoners, met with actress and peace activist Jane Fonda in July 1972.

Despite his firm, clear, and seemingly voluntary discussion in multiple interviews as a prisoner, Hoffman changed his story after he returned to tell how he was forced to make these interviews and statements and that he had been physically coerced to meet with Wald. "I reject everything I said," Hoffman said in a press conference.[33]

In response, Wald submitted his documentation of the inconsistencies to Senator J. William Fulbright, who in turn inquired of the Department of Defense. In a letter of response Jerry W. Friedheim, who provided government-funded staff support and resources to the development of Hubbell's book *P.O.W.*, said that "the circumstances surrounding interviews conducted in captivity inside North Vietnam were considerably different from those in this country when the men were free." Friedheim's letter seemed to close the matter of these inconsistencies. It seems Wald's inquiry into Hoffman's inquiry ended there. Naval Academy graduate Hoffman continued his career, eventually commanding the aircraft carrier USS *Kitty Hawk* and retiring at the rank of captain.

The most objective data we have bearing on the treatment of the POWs is the 1975 Amnesty International report that the mental and physical condition of the POWs was "really good" when they arrived at Clark Air Base in the Philippines. The report was reinforced by a 1978 study that found former POWs to be in better health than a control group of non-captive Vietnam veterans, and a later study showed them with fewer physical and psychological health problems than POWs from previous wars. Furthermore, memoirs written by POWs after post-traumatic stress disorder (PTSD) was included in the *Diagnostic and Statistical Manual* (DSM) as a diagnostic category make no mention of it and make only a few random references to trauma.

> *Wilber and Miller delivered these offerings as though they were their own original ideas; it was apparent that they believed the things they were saying.*[34]
>
> —John G. Hubbell

If a third to a half of the POWs were disillusioned with the war by 1971, the official account constructed by Hubbell contended that it was the mind-altering viciousness of the torture the downed pilots endured that made them vulnerable to North Vietnamese propaganda. The torture explanation for the dissent was complicated, however, by the presence in the Hanoi holdings, by then, of the Army and Marine ground troops and helicopter crews captured in the South. As a group, they were even more opposed to the war despite *not* having been tortured, and, as it would turn out, even less agreeable to the prisoner-at-war roles in which Stockdale and the SROs wanted to cast them.

The captives taken in the South were mostly enlisted personnel, not officers. They were younger, with less formal education than most of the pilots shot down over the North. The social distance between the SROs and the enlisted men was a good predictor of who would be a dissident. One dimension of that was the military hierarchy itself. John Young, one of the enlisted men captured in the South and eventually moved to Hanoi, later recalled that "many of the enlisted men opposed the war," a point that made sense given that a cross-section of enlistees across the military were growing restive about the war by the late 1960s and chafing under military authoritarianism. Enlisted men, moreover, had less stake in the outcome of the war than did officers whose military and political careers hinged on the success of their missions, notably senior officers, or those officers who were in a place in their careers beyond their initial commissioning obligations.

Those social status differences were intensified for the prisoner-at-war narrative by the greater contact that enlisted men had with the Vietnamese before their capture and during their days in jungle camps. Whereas officers, especially those flying off aircraft carriers—similarly Air Force officers flying from their American-style bases in Thailand—had never seen, much less interacted with, any Vietnamese before being shot down and meeting their enemy under the extraordinary circumstances of being captives. The enlisted men stationed on the ground in the South, however, are likely to have seen Vietnamese in more ordinary roles: street vendors, civilian employees on military installations, bar girls, or even girlfriends. Army soldiers would likely have been on a first-name basis with at least one Vietnamese and may have even known something about his or her family.

Captivity in the South would have exposed those Americans to the way the Vietnamese were experiencing the war. In his memoir *Black Prisoner of War*, James Daly recounts meeting the young volunteers working along the Ho Chi Minh Trail as he was moved by foot, truck, and train to the North:

> I'd seen or talked with a good number of Vietnamese. . . . They'd been peasants, Viet Cong, NVA soldiers, workers, professors, Montagnards. And the more Vietnamese I came in contact with, the more I knew in my gut . . . that the war was wrong, and we had no right to be tearing the country apart.[35]

Frank Anton, a chopper pilot downed in January 1968, recalled his treatment by an old Vietnamese woman after being captured:

> She saw my feet were wet. She removed my socks and dried my feet with a rag. She left and returned in a moment with a pair of clean white socks belonging to her. She pulled them very gently on my feet and hung mine up to dry. Other villagers had hissed

> and thrown rocks at us as we passed. But this act of kindness was unexpected and I was touched.[36]

The cross-cultural exposure that GIs and Marines brought into Hoa Lo lent an element of humanity to the Vietnamese prison workers. Like the pilots shot down over the North, the men held in the South coined nicknames for their guards but were more likely to know their real names—Huong and Qua (a Montagnard), for example—and remember others respectfully years later as Mr. Ho or Mr. Bai.

It was the humanizing of the Other that the SROs seemed to fear the most, as if recognition of an element of "us" in "them" implied the obverse: the enemy within the hearts and minds of their own rank and file. The real fear, as Craig Howes put it, was of the "white gook," the POW underling who would give in to the urges of his inner-Other and emerge as an enemy-inside-the-gates. Believing that "enlisted men collaborated [with the enemy] almost instinctively,"[37] SROs viewed with alarm any behavior that looked Vietnamese to them—squatting or eating from a bowl, for example—as signs that one of their own may be crossing over. As prophylactic to cultural contamination, SROs enforced their own segregation of POWs from the Vietnamese, prohibiting, for example, their learning of the language.

THE RAID ON THE SON TAY prison camp in late 1970 was foiled by the movement of the POWs held there into facilities closer to the Hoa Lo prison in Hanoi. That shift may have been triggered by peace negotiations and a larger effort by the prison administration to centralize the captives in preparation for the end of the war and their release. An unintended consequence of these moves, from the Vietnamese perspective, was that it brought more of the POWs under the more direct influence of the SROs who were plotting a post-release public relations campaign in which their tough-guy never-say-die personas would be featured in the prisoners-at-war narrative—the story line they now forced on their underlings as theirs as well. According to Howes, prisoners arriving from the outlying camps were fed a virtual curriculum about the bad old days during which *all* Hoa Lo prisoners had been tortured before saying what their guards wanted to hear. This became the hero-prisoner story recounted to the press and to Hubbell, which he turned into his book *P.O.W.*

Another consequence, unforeseen by the SROs, was that the unruly voices coming in from the South harmonized with dissenting voices among the pilots that were not swallowing the SRO party line. There were subtle indications throughout the years that some pilots had second thoughts about the morality of the war they were fighting and skepticism about the prisoner-at-war theatrics they had been caught up in.[38] To begin with, there were the widely acknowledged statements against the war made by even the hardest of the hardcore SROs that were then justified—or explained away—as having been coerced. And there were pilot interviews with the peace travelers that revealed misgivings about the war. Air Force Captain Larry Carrigan, for example, told the October 1967 peace group that included Tom Hayden and Vivian Rothstein that reading Felix Greene's book *Vietnam! Vietnam!* had "kinda changed my mind" about the war. Tellingly, he then asked that they not quote him on that because "they've got prisons in the States also."[39]

By the time of the POWs' release and repatriation in early 1973, the war within the walls of Hoa Lo was more than that between prisoners and their guards. The tensions between officers and enlisted men, universal in military organizations, had hardened into class lines across which officers risked the welfare of enlisted men for the sake of their own hero-captive reputations and enlisted men openly dismissed the authority of the officers. Finally, to complicate matters, the solidarity of the SROs on which the prisoner-at-war narrative rested was shaken when two of the more senior-ranking officers, Edison Miller and Gene Wilber, added their voices to the cause of peace and rebuked the chain of command that Stockdale, Risner, and Denton had contrived.

By the time the POWs landed stateside, the press was as interested in the story of dissent within the prison population as anything else. The Pentagon and the Nixon White House, on the other hand, wanted a different story told. Subsequent chapters in this book will reconstruct how those competing narratives played out in the media and political and popular culture.

2

Profiles of Dissent: Senior Officers

For his 1996 book, *The Wars We Took to Vietnam,* English professor Milton Bates studied the ways in which longstanding social conflicts within the United States played out in the war in Vietnam. Along with what he called the race, sex, and generation wars, he said that class wars rooted in domestic, occupational, and workplace social relations extended into the realities of in-country military life. "What the working-class conscripts found when they arrived in Vietnam," he wrote, "was—work."

War, Bates continued, has been compared with work, and work to war, pointing to Karl Marx's description of laborers as "privates of the industrial army . . . under the command of a perfect hierarchy of officers and sergeants," And just as workers resent the owners and managers who profit from their labor, "the worker-soldier had reason to resent those who were using him, the war's foremen and managers." Had Bates looked for evidence of the class war waged within the POW population, he would have found it there too, between senior officers and enlisted men and, unexpectedly, among the officers themselves.

LABORERS?—NOT IN THIS CLASS

POWs Edison Miller and Gene Wilber were uncharacteristic of the senior officer class. At ages thirty-six (Miller) and thirty-eight (Wilber) at time of capture, the two were among the most senior in years of service and age. They were career officers, Miller with more than eighteen years of service and Wilber with more than twenty years. They had previous combat experience, too, unlike many of the other captives: Wilber and Miller had completed air combat tours over Korea in the early 1950s. Miller flew the F-4U Corsair in support of Marine and Army ground troops and later served as a forward air controller, while Wilber flew the AD (later A-1) Skyraider on night strike missions from aircraft carriers, often landing in pitch darkness. They were combat-seasoned senior officers who had achieved fighter-squadron commanding officer roles. But their roots were lower class: they entered the military directly from high school and bypassed the normal college education prerequisite to become commissioned officers as aviators.

Despite their combat experience and advanced careers, they had developed and taken controversial stances against the war. Moreover, they were vocal and persistent in their dissent, using their rank, maturity, confidence in their experience, and perhaps even class background to disagree with the other SROs. These behaviors resulted in "stripping of rank" by the internal POW "chain of command" that the SROs claimed. Miller and Wilber were to be ignored, or worse, ostracized and shunned.

That Miller and Wilber differed from the other officers in the prisoner population by their persistent dissent coupled with their refusals to "accept amnesty"—as if the SROs had legal authority to offer it—by recanting and ceasing their dissenting actions is clearly established. It is less clear why they distinguished themselves the way they did, although social theory suggests that their upbringing, economic status, education level, and career paths, and in the case of Wilber, professed religious beliefs, set them apart from the "heroes" of the official story.

EDISON MILLER

Edison Miller was born Edward Grant Kennedy to a single mother in western Iowa on July 6, 1931. Days after his birth, she signed him over to an orphanage in Davenport in eastern Iowa. Miller describes being aware at a young age that he would have to market himself as a means to eventually get adopted, so he developed a sense of confidence and manner to win others over. The orphanage raised Edward Kennedy until Margaret Miller, a single woman attorney from Clinton, Iowa, adopted him at the age of five. With the means to support her compassion, Ms. Miller fostered and adopted ten children.

Ms. Miller gave five-year-old Edward a new name—Edison Wainwright Miller. She also gave him a role to model, as she was Miller's inspiration to become a lawyer, an occupation he practiced for nearly forty years after retiring from the military in 1973.

Miller speaks freely of his orphan years and his upbringing with a collection of adoptive siblings as having been an arena for learning to survive by communicating and promoting his own interests. As an orphan, he developed a confident extroversion as a means to secure an adopter. In a family setting after adoption, he learned to talk his way out of situations by convincing others of his viewpoint.

In his teens, Miller worked farm jobs and sold newspapers. In both work settings and school, he learned that he could enlist the help of others to get what he wanted by enticing them with food treats, typically doughnuts. Graduating from high school in Davenport, just shy of his eighteenth birthday, he enlisted in the Navy and began boot camp in July 1949. He tested highly across an array of exams and scored an IQ of 149.[1] Within his first ten months in the Navy, Seaman Miller was selected for the Naval Aviation Cadet (NAVCAD) program.

The skills that got Miller out of an orphanage and through his school years fast-tracked his Navy career. At flight school, he hung out with the schedulers, winning them over with doughnuts, coffee, and potato chips, and aligning himself with some of the Marine Corps flight instructors, who took a liking to his moxie and enthusiasm. Miller graduated ahead of his peers and, based on the urging of his flight instructors, opted for a commission as a second lieutenant in the Marine Corps, rather than an ensign rank in the Navy.

As a product of the NAVCAD program, Miller was part of a small minority in a population of college-educated aviation officers from the Naval Academy, ROTC, or Officer Candidate Schools. And having started flight training in his late teens, at age twenty he was two years younger than typical newly commissioned second lieutenants who graduated from college or the academies. In some ways, the NAVCAD path to aviator status is akin to the path to citizenship for an immigrant pursuing the American Dream: a path to great rewards potholed with the resentments of others who were humping the more "regular" paths. In casual settings such as the officers' clubs, it was not uncommon for officers to play out a kind of "caste order" of commissioning, with service academies such as Annapolis or West Point at the top, then ROTC and OCS/OTS programs, with the NAVCADs like Miller, with no college credentials, at the bottom. It was an informal pecking order with him at the low end. But it would not restrain the headstrong Miller.

Years later, he says he "did a lot of unconventional things" in his career: always willing to ask hard questions and never afraid to "challenge the system." He described himself as an iconoclastic rule-breaker: "Just tell Ed Miller the name of the game and he'll find the rules and beat you."[2]

Miller advanced in rank, gaining command of a fighter squadron that deployed to Chu Lai, South Vietnam, on July 5, 1967. From there they provided air support for ground operations below the DMZ and struck targets north of the DMZ. On October 13, 1967, north of Quang Tri Province, Miller and his radar intercept officer were hit. In a matter of seconds their plane was uncontrollable, and both ejected over hostile territory. Severely injuring several vertebrae during the ejection and breaking his ankle on parachute landing, Miller would spend more than forty days being held in field conditions before arriving in the prison facility at Hoa Lo in Hanoi.

After periods of initial interrogation and living in a cell by himself for a few weeks, Miller spent 1969 with other prisoners in Hoa Lo, and then to another detention facility, where he was joined by Gene Wilber when Wilber was moved from Hoa Lo in April 1970.

GENE WILBER

Edison Miller and Gene Wilber shared similarities in challenging upbringings and general career paths. They were from rural, farm, and working-class backgrounds and found employment and career paths in the military. In the detention system in Hanoi, the life experiences of Miller and Wilber were now set to converge.

Walter Eugene Wilber was born at home in rural north central Pennsylvania on January 17, 1930. His parents were tenant farmers, also known as sharecroppers. They worked for a landowner dairy farmer performing daily chores, milking, farm maintenance, planting, cultivating, and harvesting. This labor was in exchange for housing, produce, and a share of the profits, contingent upon the honesty of the landowner. It was a hard life; by the time Wilber was five, he could drive a team of horses. Once, helping load hay into the barn, he saw the trip-rope for the hay-carrier mechanism caught on his hayrack. With his dad beneath the load of hay, Wilber freed the rope, saving his dad from being crushed by hay. Wilber's hands bled from the rope burns so his mother slathered them with butter, wrapped them with flour sacks, and sent him out to get the cows home for milking—all in a day's work for a farmboy.

In 1942, with the Second World War underway, Wilber's parents took jobs in a factory, and were eventually able to buy a home. In the meantime, Wilber continued to work at farms. As early as age twelve, he was hired out to farmers during the summer months and received visits from his parents and siblings on Sundays. Sundays were important for Wilber, who received his "Christian education" at rural Baptist or Methodist churches, making a "decision for Christ" at thirteen. He would remain religious throughout his life, expressing protestant Christian beliefs through his identification and actions.

Wilber got a driver's license at sixteen, purchased a truck, and began driving a milk pickup route to local dairy farms, starting very early each morning and arriving at school late after delivering his truckload to the dairy plant. He maintained his studies, played basketball on the high school team, and graduated in 1947, in Troy,

Pennsylvania, four miles from the tenant farmhouse in which he had been born.

In the winter after his high school graduation, and sensing the Cold War mood enveloping the nation, Wilber enlisted in order to be a step ahead of the draft that he feared might dictate his choices. He had always wanted to learn to fly, so he enlisted in the Navy in 1948. After boot camp, schooling as an aviation electrician's mate, and getting some fleet experience, he entered the Naval Aviation Cadet (NAVCAD) program, just as Edison Miller would do the following year.[3]

Returning from deployments to Korea,[4] he continued through a progression of Navy jobs—flight instructor, flight deck officer on a carrier, jet fighter squadron pilot, and more. As a student in the ten-month graduate level program at Army Command and General Staff College, he excelled academically (his thesis graded as "superior") as the only non-college degreed officer in his class.

After a tour at Naval Air Force Atlantic Fleet as an operations officer on the admiral's staff, the young commander (thirty-five years old) had completed his professional grooming for the command pipeline. He next trained on F-4 Phantom IIs and joined Fighter Squadron 102 onboard USS *America* in early 1967 as executive officer, taking command of VF-102 on March 29, 1968, with orders to deploy the squadron with Carrier Air Wing Six to the Gulf of Tonkin.

During that period, Wilber was becoming more aware of controversies regarding conflict in Vietnam. In 1967, television news programs and the print media were reporting on protests, college unrest, and the October March on the Pentagon. At the same time, there was a reshuffling of Washington leadership surrounding the war in Vietnam that was leaving things unsettled. On November 1, 1967, Secretary of Defense Robert McNamara submitted a memorandum to President Johnson recommending de-escalation, stopping the bombing and turning over military responsibilities to the Republic of Vietnam. On November 29, McNamara submitted his resignation,[5] leaving his post on February 29, 1968. Amid that transition, Wilber was given command of a squadron with orders to go into combat. Two days later, on March 31, 1968, the commander-in-chief, President Johnson, announced abrupt changes in the war policy and "quit," announcing publicly that he would not seek reelection. The command structure far above Wilber's head had been reshuffled. On Monday morning, Wilber continued preparations for squadron deployment, and on April 10, 1968, they were underway for the western Pacific.

In the daily briefings from the Task Force Commander's staff, Wilber was confronted with rules of engagement that he annotated in his notebook as "confusing" along with reports from other pilots of operational command and control mistakes as war planners constantly adjusted to changing directions from the White House. On his twenty-first mission, on June 16, 1968, Wilber was shot down and found himself on the bank of a rice paddy west of Vinh, halfway between the DMZ and Hanoi, due to the mistaken decision of an air war commander miles away at sea.[6]

Having twisted his ankle on impact after a five-second parachute ride, Wilber was soon captured. Barefoot and blindfolded, he was led away, a young boy holding his hand and guiding him along a path. Wilber would not be heard from again for a year and a half, making an antiwar statement broadcast on Radio Hanoi in November 1969, the first confirmation for his family that he had been captured and was alive.

At the end of a nine-day trip by foot and truck from Vinh to Hanoi, Wilber suffered a stroke. Years later, the other POWs critical of his antiwar views would explain away the sincerity of his dissent by attributing it to a mix of stroke symptoms, trauma from the shootdown, and guilt for having lost his radar intercept officer—the PTSD-like explanation that later stigmatized the protests of other antiwar veterans as a mental health issue.

It's likely, however, that the stroke itself was brought on by logistical circumstances. The carrier had taken on bad water some weeks before, leaving Wilber and many shipmates weakened by dysentery. This had him dehydrated on the day he was shot down, and he was not given much to drink on the steaming-hot trip to Hanoi. Beckoned from the back of the truck at the entrance to Hoa Lo prison, Wilber could not stand. Carried inside, he could only sit on a stool if they propped it in a corner.

MILLER AND WILBER IN HANOI PRISONS

Miller was in Hoa Lo prison when Wilber arrived in late June of 1968, but they would not meet until April 1970. Wilber's first twenty-one months were spent living alone without contact with other Americans at Hoa Lo. The POW literature has used the term solitary confinement,[7] which might sound more ominous than it was. Some, but not all, new shootdowns were put in solitary, and Ralph Gaither, in his memoir *With God in a POW Camp,* said no one understood why.[8] Prison camp officials have said the reason was due to space. When there were more rooms than prisoners, nearly all were given single rooms, that is, put in solitary; as populations increased, prisoners were given roommates. Heard by Americans, solitary confinement connotes dungeon-like conditions, perhaps even a torture chamber—and that's the way Hubbell and Rochester and Kiley leave it. When pressed on it in a later interview, Wilber said his incarceration was "not very kind treatment" but that he "was not physically beaten or—or handled in this way, which we've heard described before."[9]

Released from solitary in April 1970, Wilber was moved from Hoa Lo to the facility the prisoners called the Zoo. About four kilometers southwest of Hoa Lo, the Zoo was originally a French film studio, Pha Phim.[10]

At the Zoo, Wilber would meet his first roommates, Miller and Commander Bob Schweitzer. Like Miller and Wilber, Schweitzer was a NAVCAD, having passed on college as an enlisted trainee, earning his officer's commission when he earned his aviator's wings.

Miller and Schweitzer had been exchanging views critical of the war before Wilber arrived. They were already somewhat notorious among the other pilots for a tape recording that had been played over the camps' public address systems; the other inmates derisively referred to the broadcast as "The Bob and Ed Show."

Bob: I think you're right, Ed. We have no business being in this war. It is strictly an affair for the Vietnamese to settle among themselves.

Ed: Not only do we have no business being here, our interference is illegal. It's an undeclared war of aggression, Bob, and that makes it downright criminal. That makes us criminals.

Bob: Actually, the Vietnamese government has every right to try us as war criminals. We have forfeited any rights under the Geneva Convention on the treatment of POWs. Ed, even the Code of Conduct has no application here.[11]

In their room at the Zoo, Wilber joined Miller and Schweitzer in discussions, sharing his thoughts about the impropriety of the war that he had come to realize during his time in solitary. On occasions, the three of them met together with visiting international delegations, but their most important exercise of dissent was through Canadian television journalist Michael Maclear.

THE MACLEAR TELEVISION INTERVIEW

Maclear was the only Western television journalist allowed into the Democratic Republic of Vietnam during the American war with Vietnam. He had reported on the funeral of Ho Chi Minh in September 1969, bringing news footage to American televisions. On his next trip, in December 1970, Maclear was taken to the Zoo on Christmas morning. He shot silent footage of the facilities, showing pilots Paul Brown, Mark Gartley, Edison Miller, Roger Ingvalson, Gene Wilber, Bob Schweitzer, and William Mayhew in the foreground. Maclear filmed a twenty-minute interview with Schweitzer and Wilber. Maclear's questions had been submitted to Wilber and Schweitzer ahead of time, and some of their responses had been approved by prison staff.

Maclear's film was developed in Tokyo and transmitted by satellite for an evening-news special on NBC and CBS on December 27. The CBS broadcast was introduced by Charles Collingwood, a network regular at the time. Acceding to the Cold War climate, Collingwood signaled to viewers that what they were about to see and hear was a Communist propaganda stunt in which U.S. prisoners of war were being used as props. The "plight" of the prisoners had become an issue in the negotiations to end the war, he said, and warned that the North Vietnamese authorities in Hanoi "were anxious to put their contention [*sic*] that all American prisoners of war in North Vietnam are well treated." What they were about to see, Collingwood advised, was "a carefully controlled Christmas interview with carefully selected American prisoners."[12]

Maclear then narrates what viewers are seeing as his camera pans across the camp and prisoners' living quarters:

> The rooms, despite bars and bolts, could hardly be called cells.
>
> Twenty feet wide by twelve, each had three beds but only two to a room were made up, two blankets on each.
>
> Family pictures adorned and warmed some rooms.
>
> Books were generally evident, and some newly arrived gum and gifts. But there are not many normal everyday articles, like ashtrays, pencil and paper or items of clothing.

Then the interview begins with Wilber and Schweitzer introducing themselves. With a Christmas tree in the background, they spoke without notes and did not appear to have been rehearsed. The format of the interview was straightforward, with brief to-the-point answers to each question. Maclear began with questions about the camp:

Maclear: Wilber and Schweitzer spoke easily, articulately, and with no trace of embarrassment. Some comments of no apparent significance were later censored.

What letters and parcels do you regularly receive, what do the parcels contain in them from your people, and what letters do you send?

Wilber: We get letters about every month, regularly about once a month, packages about every two months, and my packages usually contain candy and various food items, that uh, special little snacks, you know like peanuts and things like that, and sometimes some underwear, and small items. Chocolate candy and things we appreciate all the time, chewing gum.

Schweitzer: And, of course, our wives send the usual underwear, handkerchiefs, socks. We don't really need any clothing, but you know how wives are.

Maclear: And what mail do you send?

Schweitzer: We regularly send one letter a month. It's a regular form letter that both our families use and we use. It's been arranged by the Committee for Liaison, I believe, and the Vietnamese American Friendship Association. And I believe they go through Moscow. Through the new mail service, I believe from Moscow to New York and so on.

Wilber: Also, we occasionally send long letters for Christmas, Mother's Day, special occasions . . . we make radio messages for our children's birthdays, and our parents' birthdays, and our wives' birthdays. So we make many radio messages each year, many.[13]

For about a minute, Wilber and Schweitzer provide other details about the conditions such as meals, the books and magazines they had been given, and describe the events in Hanoi that they have been allowed to attend such as a Russian film production of Shakespeare's *Twelfth Night*, after which they were given an English edition of *The Complete Works of Shakespeare*. Then the interview turns more political:

Maclear: I want to ask do you talk to each other about war, what are your feelings on this, and what might you want to say directly to the American people?

Wilber: Yes, we discuss the war very much because the war is very close to us here. And we are all involved with it.

Schweitzer: I know I've had the deepest discussions I've ever had in my life with my fellow prisoners here. And we've had to go to the very core of our feelings on a number of things—loyalty, what is it, where does it lie; and morality, legality, a number of things that in our affluent, rushed life, I suppose in our country we don't normally give too deep thought about. But here we definitely do. I feel all of us do. And we've talked about it at great length, and how we feel.

Wilber: I think the answer of course is that the war must be ended, and it must be stopped now. We've just got to stop this thing. We've got to admit the facts as they lie and stop the war, and of course we must withdraw our troops to stop the war. That's a condition we have to face. So, once we do that the Vietnamese can solve their own problems, I'm confident of that. If we'll stop the war and get our troops out, that's what we have to do, and that's what the big job is.

Schweitzer: I of course agree, and, as I say, I'm terribly concerned about my country and I feel that the future of our country as well as Vietnam and Indochina cannot be served by the prolongation of this war, whatever the reasons and causes. I don't feel that it's necessary even to rake over the old reasons of who was wrong, who was right. It has been proven as far as I'm concerned. This war is bad, is bad, and it isn't going to improve, either our situation or the Vietnamese or Indochinese peoples' situation, they've got to be left alone.

Maclear: I want to thank you both. I think what you've had to say can only but help.

Collingwood then provided a summation for the program that avoided the message sent by Wilber and Schweitzer—that U.S. government and press claims of their mistreatment and subjection to mind control was anti-communist propaganda—and attacked the messengers. "Obviously," he said, "the North Vietnamese would not bring forward either recalcitrant or physically debilitated prisoners to be interviewed." The films and books that Wilber and Schweitzer had access to had "been restricted to propaganda and antiwar books, as is perhaps natural except for the works of Shakespeare which presumably are sanitized by now. . . . We can only hope all the other Americans prisoners of war in Vietnam had also been as well treated as those we have heard. That is obviously what the North Vietnamese want us to think."

Broadcast by NBC and CBS as a special report, at a time when American had only three network choices for television news and entertainment, ABC being the third, Maclear's report, with Collingwood's spin, could hardly have reached a wider audience. Attention to it was then amplified in the news cycle that followed, which continued to discount the message of Schweitzer and Wilber. The December 29 *New York Times* headlined, "Laird Discounts P.O.W. Interviews," referring to Melvin Laird, the secretary of defense, while the *Times*'s special correspondent, Christopher Lydon, reported on December 28 that a Pentagon spokesman declared the camp shown by Maclear to be a "showplace."[14] White House press secretary Ron Ziegler quoted President Nixon's comment that the North Vietnamese facilitation of the interview was a "total disregard of the terms and intentions of the Geneva Conventions."

AFTER MACLEAR

The thumbs-down press response to the Wilber and Schweitzer testimonies might have discouraged the POW dissenters. After all, if the public could not be rallied to the courage of their heartfelt pleas for peace, what chance was there that the war could be ended, and that they would be spared social and legal reproach upon return, if and when that came to pass? Nevertheless, the rebel captives stayed the course, producing, in Hubbell's words, "all kinds of the most virulent antiwar even anti-American, material for the enemy." As an example, Hubbell quoted Ed Miller's 1971 Mother's Day message:

> Today, America's mothers must face the fact that their sons are killing fellow human beings and destroying foreign countries for

> an unjust cause, making our actions not only illegal, but immoral. . . . My personal participation in this war is my personal shame and tragedy. My country's immoral and illegal actions which are now culminated in the tragedy of Vietnam is America's shame.[15]

According to Hubbell, on July 4 Miller and Wilber tape-recorded an open letter to the American people in which they described themselves as patriotic officers dedicated to their country who had been deceived by their leaders into committing acts against Indochina that were immoral and barbaric. They ended with an Independence Day wish for the success of the antiwar movement.

Changes were afoot, however. At the same time Maclear was in Hanoi, the Vietnamese were closing smaller camps such as the Zoo and consolidating prisoners into the Hanoi city-center prison, Hoa Lo. The move was prompted by the U.S. special forces raid on the Son Tay camp just outside Hanoi the previous November, and by the prison authorities' desire to manage more closely the prison population as negotiations to end the war proceeded.

Meanwhile, the Defense Department and press reaction to the Maclear interview had prefigured a Pentagon counterstrike against POW dissent in Hanoi. Coincidentally, the crackdown was simplified with Wilber, Miller, and their sympathizers now consolidated under the thumb of the hardline SROs in Hoa Lo. In midsummer 1971, the Navy's James Stockdale ordered Wilber, Miller, and Schweitzer to "Write Nothing for the V. Meet no delegations. Make no tapes. No early releases."[16] He finished by demanding to know from them: "Are you with us?"

Schweitzer responded to Stockdale's order equivocally, saying "each man should make his own decision" but that he was "in general agreement with Commander Stockdale."[17] In August, he agreed to amnesty from prosecution that was threatened by the SROs and rejoined the SRO-designed prisoner command structure. In return, he agreed not to make any more statements nor participate in any more interviews.

Wilber and Miller responded forthrightly: "We actively oppose this war." Wilber's commitment to peace was a "matter of conscience" supported by his deeply held religious beliefs. He had concluded while in solitary confinement that the war was not declared legally, and that an order forbidding him to express his opinion was not a "lawful" order under the Uniform Code of Military Justice.[18] He believed that his First Amendment rights as a citizen prevailed, and that the dynamic nature of the military required a robust discussion in order to get to the best decisions.

On August 11, after being offered amnesty in exchange for their refusal to desist from public objections to the war—the offer that Schweitzer had accepted—Robinson Risner advised Wilber and Miller, "You are hereby relieved of your military authority."[19]

In their last year of captivity, Miller and Wilber gave more interviews and made several more statements. In May 1972, with six other POWs, they spoke with a team of peace activists that included Clergy and Laity Concerned's Reverend Robert Lecky, Catholic priest Father Paul Meyer, president of the National Student Association Margery Tabankin, and Bill Zimmerman representing Medical Aid to Indochina; in July along with five other POWs they met with actress and activist Jane Fonda, and in August with former U.S. attorney general Ramsey Clark.[20]

The so-called Christmas Bombings of 1972 resulted in ninety-two American fliers being shot down (thirty-three KIA, thirty-three POW, twenty-six rescued); the prison population swelled with thirty-three more captured pilots being brought into Hoa Lo. Many of the new captives were young junior officers who were warned through the prisoner grapevine to steer clear of Miller and Wilber, as new captives had been warned for some years by then. Years later, however, one of the newer shootdowns remembered there having been something compelling about these two senior officers who defied convention and knowingly risked their careers despite threats of legal punishment.[21]

Together with Wilber and Miller, six of the newbies shot down in 1971 and early 1972 signed letters decrying the "many innocent people dying a totally needless and senseless death" in the bombings.

The letters were addressed to senators and representatives and media leaders, including newscaster Walter Cronkite.[22]

RELEASE, RETURN, REJECTION

Threats by SROs to punish the pilots who had spoken out against the war, and the anxiety of the dissident pilots about trials, ruined careers, and prison time, were put in abeyance by Secretary of Defense Melvin Laird's declaration that alleged offenses by POWs were to be forgiven.[23]

Had Wilber and Miller gone to trial, it is likely their defenses would have taken very different forms. Miller, characterized by his fellow prisoners as "a lawyer," would argue there was insufficient evidence to support the charges. Wilber, on the other hand, didn't deny the substance of the charge that he had made statements opposing the war, but he remained convinced that the war was wrong and that his objections to it were morally justified. In late March, Wilber was contacted by Mike Wallace for an interview on the CBS *60 Minutes* show that was filmed on April 1 and aired the same evening. If Wilber had thought Wallace would reprise Maclear's straightforward questions about the conditions of his imprisonment, he would be surprised. Wallace set up Wilber for a "brainwash" indictment in his opening monologue:[24]

> Tonight, a POW, a Navy captain who says he was not tortured in Hanoi, a man who made antiwar statements broadcast by Radio Hanoi. . . . Wilber says he made his antiwar statements in Hanoi voluntarily. More on that story from Captain Wilber when *60 Minutes* continues.

Not so subtly, Wallace was cueing his viewers that Wilber had not sought relief from physical pain by making the statements against the war, an explanation that all Americans could understand and sympathize with, but that he had fallen prey to communist mind-control tactics. He had been *persuaded* to make his statements, not *coerced*.

Without ever actually alleging that Wilber had been programmed to renounce his loyalty to America, Wallace had cleverly trapped the captain in a kind of damned-if-you-do, damned-if-you-don't conundrum: Wilber could either contradict his previous statements about not having been tortured—and thereby destroy his own credibility and demonize the Vietnamese—or he could deny, as Miller was doing, that he had made the statements at all.

Wallace seemed most intent on forcing Wilber to denounce the Vietnamese for torture and, failing that, impugn Wilber's character.

Wallace: Captain Wilber, we've heard so many awful tales of torture of our POWs this week, all week long. What was your experience?

Wilber: Well, I was not tortured.

Wallace: Surely you have no reason to disbelieve the stories that we're hearing, then, from your fellow officers.

Wilber: No, sir, I would not disbelieve them or attempt to repudiate them in any way. It's something that each person has to tell his own story.

Wallace: Someplace along the line in prison camp—maybe it was while you were in solitary—you apparently changed your mind about the war. After all, you were a career officer—you weren't a child. You were already in your late thirties.

Wilber: Uh-huh.

Wallace: What changed your mind about the war? What made you—

Wilber: Well, I think, my conscience. I happen to be a Christian and found out that my conscience bothered me. And so, the—the big old bugaboo that we hear—conscience and morality—started to show itself. We have to look at things in, in the total picture of legality and morality. And as I had time to sit for many hours . . . I had time to really find out what Gene Wilber was like.

Wallace: When your fellow POWs learned that you were making antiwar statements—and you did, more than once—meeting with American antiwar groups in Hanoi, and so forth, what was their reaction to you?

Wilber: Well, it was varied, and this is not unusual. Some concurred, some encouraged it, some said, "I'm with you, but don't include me." There—it was just a whole range of feelings. . . .

Wallace: Well—do you know Captain James Mulligan?

Wilber: I know him very briefly.

Wallace: The reason I ask is that he, another naval officer, another POW, is quoted in the *New York Times* just this morning as saying, "No one ever minds anyone breaking under torture. But it's those guys who fink out who get you. Those guys will get what's coming to them."

Moments later, Wallace returned to Mulligan's threat that Wilber and other dissidents "will get what's coming to them." Some of those guys, Wallace repeated, "They say that they're—they're going to get you. They say they're going to—they're going to follow this through. There's talk of court-martial, there's talk of ostracism." Wallace had gone out of his way to quote Mulligan's words from the *Times* story, and his reiteration of "get you" hung in the air as the interview ended. Had he gotten wind of plans for violence against Wilber, Miller, and others? By alluding to its possibility, had he, if inadvertently, encouraged just that?

Howes in fact wrote in *Voices of the Vietnam POWs* that one SRO in April 1972 had issued a "conditional license to kill" fellow POWs if their loyalty to the United States was suspect. The circumstances surrounding the threat will be clarified in chapter 5 of this book, but the likelihood that Wallace was privy to the story, otherwise unreported, raises questions about who leaked it to him and for what purpose other than to unnerve Wilber and provoke real harm to him and the other dissidents.

The prescience of Wallace's allusion to violence was soon demonstrated. On May 24, President Nixon hosted the POWs, spouses, and dates for a White House reception. It was a gala affair with more than 500 former POWs in service dress uniforms with spouses and dates in evening dress. Earlier in the afternoon, after a briefing of Navy former POWs with the CNO, Admiral Zumwalt, Wilber and his wife returned to their hotel to prepare for the evening dinner events, only to find out that their room had been broken into. The word “TRAITOR!” was written in lipstick across the mirror, among a few other epithets; a fishbowl with goldfish had been placed in the room (evoking the trope that antiwar POWs had an aquarium as a part of alleged special treatment by their captors); and personal items were in disarray.

The Wilber hotel room break-in was the kind of calling-card operation that sent the message: “We can get you when we want.” In the paranoid climate surrounding the Watergate break-in that was bringing down the Nixon presidency, it was not an uncommon tactic. But Wilber did not report the incident, and it proved to be a one-off event.

The threat of courts-martial and ostracism, also alluded to by Wallace, were more fully realized. When Secretary of Defense Laird declined to bring Defense Department charges against the dissidents, he left open the possibility that *individual* POWs were free to file charges against other individuals, and that's what Rear Admiral James Stockdale did. In late June 1973, he charged Wilber and Miller with “mutiny, aiding the enemy, conspiracy, soliciting other prisoners to violate the Code of Conduct, and causing or attempting to cause insubordination and disloyalty.”[25]

A three-month investigation by Navy Secretary John W. Warner dismissed the charges, saying courts-martial would be disruptive to those who would have to testify. In September 1973 he issued letters of censure to Wilber and Miller, charging them with failing to meet the standards expected of officers of the armed forces and announced they would be retired in “the best interests of the naval service.”[26]

The ostracism foreseen in the Wallace interview also loomed. Ad hominem humiliations such as Wallace's infantilizing remark that Wilber was “not a child” would dovetail with the SROs' attempt to distinguish their upright manliness as hardcore loyalists from the weaklings who had sought relief from torture in return for giving aid and comfort to the enemy. The projection of weakness onto fellow prisoners like Wilber and Miller, who were less educated and less privileged by class background, would be most apparent in the attitude shown by SROs toward enlisted men who returned with them—the subject of the next chapter.

Ostracizing of that type was ingrained in the notion of brainwashing that was carried from the post–Korean War studies of POW defection into the years of the war in Vietnam. The psychologizing of dissent characteristic of those studies became, in turn, the backstory to the medicalizing of GI, veteran, and POW rejection of war, discussed in subsequent chapters.

3

Profiles of Dissent: "The Peace Committee" of Enlisted POWs

The enlisted status of those in the Peace Committee made damning them relatively easy.

—Craig Howes

The media spotlight on the pilots returning from Hanoi in 1973 overshadowed the news about returning ground troops.[1] News organizations were far more interested in a few pilots like Ed Miller and Gene Wilber accused of collaboration than in the men captured in the South and eventually moved to the Hanoi prison facilities. Most of the latter group were Marines and Army soldiers. Few were officers. Among them were eight enlisted men known as the Peace Committee.[2]

Charges were brought against "The Eight" by Air Force Colonel Theodore Guy. While not from the same background of service academia as James Stockdale, who charged the dissident pilots, Guy's pedigree was nevertheless privileged in comparison with the men he accused. His father was a professional musician who had traveled with the Dorsey Brothers band. Guy had attended Kemper Military School in Missouri and then was wait-listed for admission to West

Point. Impatient and wanting to get his military career underway, he joined the Air Force in 1949. Over Korea, he won two Distinguished Flying Crosses and six Air Medals.[3]

Prior to his capture in Laos, Guy had also known a war in Vietnam that was far removed from that of the men he would accuse. The night before getting shot down, he had enjoyed a bucket of popcorn and a couple of drinks in his "sleeping trailer" after returning from another mission to his F-4 base at Cam Ranh Bay. In the morning, he had a breakfast of scrambled eggs and toast "at the club" before taking off and being shot down later in the day. Guy was captured on March 22, 1968. His back-seater did not survive the shootdown.[4]

On May 29, 1973, Colonel Guy charged the eight enlisted men with "aiding and abetting the enemy, accepting gratuities, and taking part in a conspiracy against the U.S. government." As characterized by the historians Stuart Rochester and Fredrick Kiley in their 1998 book *Honor Bound: American Prisoners of War in Southeast Asia, 1961–1973*, a reiteration of John Hubbell's *P.O.W.*, the men in Guy's crosshairs were:

> soldiers and Marines "captured in . . . South Vietnam . . . young enlistees, "grunts," average guys from ordinary backgrounds suddenly thrust into extraordinary circumstances [and] victimized as much by their youth and inexperience as by the South's peculiarly adverse conditions. The dozen or so . . . averaged barely 20 years of age, 12 years less than the aviator-officers incarcerated. . . . Their lack of training and PW organization inherent in the Southern situation, left them especially ill-equipped to deal with the psychological demands of captivity. . . . As a group they tended to be less stable and steadfast and coped less successfully than older, more seasoned hands.[5]

Drawn by Rochester and Kiley, this profile of the dissenting enlisted men was a composite of the captives' demographic details—their young ages, for example—and the authors' conformity to what had become, by the time they wrote in the 1990s, the prevailing condescension that enlisted men's own shortcomings had betrayed them. Had Rochester and Kiley drilled more deeply into the available material on The Eight, they might have found their dissent rooted in the same structure of class hierarchy as that of the antiwar pilots. James Daly's biography makes the point.

JAMES DALY

The strongest case that dissent was motivated by conscience is made by James Daly in his 1975 memoir written with Lee Bergman. Rochester and Kiley knew about the book and used it to document some details of camp life of the prisoners taken in the South, but nowhere in their *Honor Bound* did the authors use the memoir to profile Daly, the person.

Daly arrived in Chu Lai, South Vietnam, in October 1967 and was immediately moved, emotionally, by the despair of the Vietnamese. "I really felt sorry for [the kids]," he wrote, "skinny, torn shirts, bare feet, big sad eyes just staring at you." He also felt the Vietnamese "hatred for the Americans," based, he thought, "on the misery of [the] peasants and farmers [having] suffered through so many years of war."[6]

Assigned to Alpha Company of the 196th Light Infantry Brigade, Daly was wounded and captured by North Vietnamese soldiers after a fierce firefight on January 9, 1968, in the Que Son Valley west of Tam Ky. He was handed over to forces of the NLF and moved from camp to camp with other captives known later as the Kushner group.[7] Taken 500 miles by foot, truck, and train to Hanoi, the group arrived in April 1971 at a prison in Hanoi about a kilometer north of the Hoa Lo prison known by the Americans as Plantation.[8] There, they learned about the Peace Committee (PC), a group of POWs said to be sympathetic to the North Vietnamese; one of the prison guards showed the newcomers a magazine published by the PCs called *New Life*. The PCs, said the guard, "believe the war is wrong and are helping to bring it to an end."[9]

Daly's initial dissent took form in discussions with roommates about whether to write letters home expressing antiwar views. "As it turned out," he wrote in his memoir, "all the guys in the room agreed to write, and probably most everyone in the camp did, also." He addressed one letter to the American people and one to the mayor of New York City. In his letters, he "denounced the war and questioned why millions were being spent to bomb the Indochinese people, and how this money was needed at home."[10]

At Christmastime 1971, Daly and roommates were allowed to mingle for the first time with the PCs. By then, he was reflecting on the Vietnamese people he had met in the South, the troops heading south on the Trail, and the prison staff at Plantation. He thought about the history of Vietnam's struggle against European colonialism, which he had learned about reading the periodical *Vietnamese Studies*, Felix Greene's *The Enemy*, and *The Pentagon Papers*, all furnished by the guards. As he would recall later, "I knew in my gut *what I had believed before in my head*—that the war was wrong, and we had no right to be there tearing this country to pieces." When the PCs invited Daly to join them and move to their quarters, he accepted the offer.[11]

Daly's Vietnam experience might have precipitated his coming to conscience, but his words, italicized above, belie the singularity of his experience as the difference-maker in his turn against the war. Rather, it was the chemistry of his time at war and in captivity with his growing-up experience that was epiphanous.[12]

Daly was Black. He grew up as a Baptist in the Bedford-Stuyvesant section of Brooklyn. As a twelve-year-old, he hardly ever missed a Sunday church service. He really liked the church, especially "joining the different clubs and going on trips." In 1959, "a lady selling magazines" introduced the family to the Jehovah's Witness religion. His mother became a member. James did not because it involved a commitment to 100 hours a week as a Pioneer going door-to-door to spread the faith. He thought it more important to get a job to help his mother, who struggled to support seven kids. But he learned and understood the principles of the religion and subscribed to its teachings on war.[13]

Daly's father was strict and never played with him or encouraged him to take part in sports. His father left the family while James was still in school. Mostly, he wrote, he stayed in the house and helped with the cooking and cleaning. He took odd jobs "cleaning houses, washing windows or cars, and even baking cakes and pies on the holidays."[14] The jobs earned him only five dollars a week, but with that he could buy things his mother could not afford, like clothes, and still give some money to her.

Graduating from Franklin K. Lane High School in June 1966 with plans to attend community college, James took a job clerking at B. Altman's department store. However, U.S. involvement in Vietnam was deepening and he felt the draft breathing down his neck. Upon visiting an Army recruitment center, he was told, "As a conscientious objector you can pick the kind of noncombat job you want. After basic training, the choice is yours."[15]

Basic training was hard. With few physical skills, he was remanded to extra pushups. He thought the marching cadences like "Jody's got your girl and gone—Jody's got your mama, too" were insulting. But he kept his mouth shut. When he didn't yell "Kill!" on the bayonet course and explained to the sergeant that he didn't believe in killing, the sergeant called him a "pussy."[16] Through it all, James believed that a noncombat job awaited him at the end of basic training. After all, that's what the recruiter had promised. Upon completion of basic, however, he learned that the guarantee was good only for those who had enlisted for four years. James was a draftee with a two-year obligation, and was headed now to Fort Polk in Louisiana for advanced infantry training.

Daly's weeks at Fort Polk were filled with appointments, phone calls, and paper filings in pursuit of the conscientious objector status he desired. He had a "mental hygiene consultation" with a psychiatrist, meetings with two different chaplains and the commander of the 3rd Training Brigade, who suggested his best option: "Go home on leave—and stay home! Let the MPs pick you up. You'll be charged with AWOL, but you'll get attention and be able to plead your case."

At home on leave before shipment to Vietnam, Daly looked into flights to Sweden where draft resisters and deserters were finding sanctuary. He consulted the "head overseer" of his family's Jehovah's Witness congregation to see about an endorsement for his application for CO status. The overseer declined to help, upbraiding James for not accepting a Pioneer mission that would have credentialed his religious claim as a CO. Nor could the overseer countenance desertion: "You are in the service now and you must go by the rules and regulations of the military."[17]

Daly made one last appeal, appearing in person at Fort Hamilton in Brooklyn where he was told to wait until he arrived at Oakland Army Terminal in California where he would be processed for Vietnam—and to raise the issue there. In Oakland he was told again to wait, and to take up the matter with his new unit when he got to Vietnam.

Daly was in Vietnam only about three months before he was captured. From that point on, his biography as a dissident POW begins paralleling the lives of his newly made comrades-in-captivity. With remarkable sociological intuition, Daly seems to have sensed the class backgrounds common to PCs as a basis for mutuality and the strengthening of his own commitments to conscience. In the group of eight enlisted men eventually charged for misconduct, that affinity was felt first and most deeply with Bob Chenoweth.

BOB CHENOWETH

Daly did not meet Chenoweth until Christmas 1971 when the prisoner groups segregated at Plantation were freed to socialize together. Daly recognized him immediately as the leader of the Peace Committee. "There was never any question about his sincerity, involvement, or understanding of the PC goals."[18]

Chenoweth had joined the Army in 1966 and was trained to work on UH-1 Huey helicopters. He arrived in Vietnam in January 1967 and began flying combat, medevac, and resupply missions as a crew chief. On February 8, 1968, he was flying back to Da Nang in the beat-up Huey, with a black cat painted on its nose, when it came under heavy ground fire and crashed in a cemetery.[19] All six men on board got out, but they were quickly hemmed in by local NLF forces, and surrendered. Chenoweth was twenty and would spend the next five years in enemy hands.

Chenoweth had been in Vietnam for only a year. But like Daly, he had developed an empathy for the Vietnamese people and a distaste for the racist views of most Americans toward the Vietnamese. In a 2017 interview, Chenoweth reflected that he couldn't see how U.S. forces could possibly be helping the Vietnamese given the attitude that GIs had, viewing them as "subhuman" and disparaging them as "gooks and dinks."

Chenoweth was born in Portland, Oregon, December 31, 1947. His dad was a telephone company technician and a Second World War Navy vet, having served on submarines out of Hawaii. He remembers:

> The family rented houses and apartments and lived from paycheck to paycheck. When we got a little older, we picked beans, berries and nuts at the local farms. We would get to keep a little money for a toy, but most went to Mom. I don't remember a time when I had my own bedroom. I did not mind or think it unusual since I had little to compare my life to anyone else.
>
> I grew up believing America was the best place to live. I watched John Wayne and all the rest. I started building model airplanes when I was a little kid and still make models today. We played with neighbor kids including John whose family was Mexican and owned a Mexican Restaurant. John's mom was very sweet and fed us neighbor kids tacos. It was my first exposure to another culture.

Chenoweth grew up with racial sensitivities, his father being from the North and his mother from the South, with "that open outspoken self-assurance that whites were superior." In Portland he was "exposed to Black people, as well as Japanese and Chinese kids," enabling the development of his own sense of those he would befriend. At Portland's Benson Polytechnic School he was a drummer in the band and became best friends with Black kids, helping them to join the Drum and Bugle Corps to which he belonged. Chenoweth remembers a "great civics teacher in his senior year" who really helped him understand the civil rights struggle. But he says:

> It was the Army experience in both Louisiana and Alabama that shocked me into the reality of race. I pulled a lot of KP in basic training and almost always did so with Black soldiers. We talked a lot since I was curious about their lives especially in the south. One guy I met had been a high school science teacher in Mississippi before being drafted. He said there was a Black HS and a white HS in his town. He talked about the differences in equipment, classes and textbooks. What he was telling me outraged me since it went against what I believed it was like to be Americans. . . . The poverty I witnessed was also a bit of a surprise. Just to see it. Again, I did not understand about "institutional racism," etc. but I remembered what I saw. I found Army training to be the foundation of the race hatred soldiers later exhibited toward the Vietnamese.

When Chenoweth got to his helicopter unit in Vietnam in January 1967 a Black crew chief became a mentor. He introduced Chenoweth to "Ti Ti Boy, a Vietnamese guy who bought cigarettes, soap, records and other PX items. Ti Ti Boy hated Nguyen Van Thieu, Nguyen Cao Ky, and all the rest of the 'General Presidents.' I also witnessed press gangs rounding up young men for the South Vietnamese Army (ARVN), often under great duress."

The contact that ground troops had with Vietnamese before they were captured distinguished them from the pilots who had had no, or little, exposure to Vietnam and its people prior to being shot down. And that difference continued into their prison experience. Chenoweth remembered the work crews he was assigned to at the Portholes camp in Nghe Anh Province:

> [At the] the camp in Nghe An we went out of our rooms most days to repair bomb shelters, do garden work (planted watermelons, other melons, cassava, peanuts, greens, etc.), and to help in the rice fields. . . . We also went out sometimes in the early evening to go to another village and get supplies. We carried rice, blankets, giant woks, cooking tools, sugar, POW cloths etc.

> which we would bring back to our camp and others. . . . When we went to get supplies, mostly people were kind to us and curious . . . the supplies were for us, the villagers and the guards, and our camp.
>
> In Ha Tay Province our outdoor work was cleaning and gardening. We planted and tended to various plants including kohlrabi, bok choy, onions, and other greens. Once on a run, this French guy, Goiun, was carrying a huge wok. He had it over his back holding each handle in his hands. I was behind him [when] this old lady jumped out from behind a bush and wacked him with a bamboo stick. Goiun moaned and dropped. He dropped the wok and it made a thud. Duc [the village turnkey] ran up and got hold of her, pleading, "Grandmother, grandmother." Even at the time I thought it was one of the funniest things I had ever seen.

Chenoweth recalls "falling in" with King David Rayford Jr. and Alfonso Riate "just because of what I had experienced."

KING DAVID RAYFORD

Chenoweth recalled that the Peace Committee germinated with him, Rayford, and Riate while they were held at the Portholes prison camp in Nghe Anh province near the North Vietnamese city of Vinh. It was a relief for some prisoners to be there, better sheltered, fed, and treated for injuries than they had been in the jungle camps in the South.[20]

Rayford had been captured in July 1967 and was merged, en route, into the "Purcell bunch" that included Chenoweth, arriving at Portholes in April 1968.

> Ray [Rayford] and I were next to each other at Portholes where he told me his background and upbringing. He was a 20-year-old African American drafted off the Ford assembly line in Detroit. He came from sharecroppers in Mississippi as I recall. His

> grandma sent him up to Chicago. Ray had worked, had a little boy, and been injured working on an upholstery machine at a company that made barber chairs. His hand was damaged, and he had two stiff fingers. I found it incredible that he had been drafted and accepted into the Army. Ray did too. He was one of the most honest and hardworking people I had ever met. When we went out to work [from the Portholes camp] we often got pieces of pineapple, or peanuts from the villagers.[21]

It was with Rayford, in the Portholes camp, that the Peace Committee began to take shape. They were soon joined in the effort by Riate.

ALFONSO RIATE

If the Damned Eight story was to be made into a film, the trailer for it would feature Alfonso Riate. Riate was born in Sebastopol, California. His dad was Filipino and died when Al was very young. His mom was Native American from the Karuk tribe. She was famous as a tribal elder and teacher of the tribe's language. Riate's granddad practiced Indian traditional medicine. Al and his younger brother John eventually moved to Los Angeles where they lived alone. Al forged papers so they could go to high school.[22]

Chenoweth remembered meeting Riate at Portholes:

> I both admired and respected him right from the start. We got to that camp sometime in April 1968. The camp was a microcosm of life back home. When our group came [into Portholes], including Rayford, we quickly divided up along "race lines" (for lack of a better term). I had a long hatred for racist bullshit that went back to my high school days. Rayford and I talked about race, slavery, and the like since we were next to each other.
>
> I can tell you that Al had a very hard early life, but I know it was not a life without love. Reservation life, and all its hardship

> and injustice and racism, is not widely understood by "white America," even today. In the 50s and 60s it was brutal. Al's decision to leave the reservation for LA with his brother showed both strength and courage.
>
> After high school, Riate took classes at Long Beach Community College and joined the Marines in 1965. He volunteered for duty in Vietnam and went to K company, 3rd Battalion, 3rd Marine Division (K-3-3). He was captured while on an operation near Hill 861 in Quang Tri Province.
>
> Al once tried to escape from Portholes. His Filipino-Indian ancestry gave him a dark complexion—and the best chance that any of us could be mistaken for a Vietnamese. He had also learned a little Vietnamese language. On a work detail out of the camp, he escaped. Two days later, he walked into an NVA anti-aircraft site, sat down for tea with the soldiers! One of [the NVA] went off shortly after that and after a while the camp guards came down the trail to take him back to camp.

Chenoweth's lesson from Riate's attempted getaway was "You can get out of the camp but not out of the country!"

In a screenwriter's hands, the character of a mixed-race Marine who walked onto an enemy firebase and sat for tea might get embellished as a writer, soldier-poet, and singer-songwriter. But Riate was all of those, too. While captive, he wrote a protest song in Vietnamese and later recorded it.[23]

In April 1972, U.S. B-52 bombers rolled in on Hanoi. In his memoir, Daly remembered feeling close to the Vietnamese. "All I could picture," he wrote, "was some man or woman or little kid being shattered to pieces by an American bomb."[24] During the raid, Daly remembered, Riate had started writing a letter to the prison administrator. When the bombing stopped, Daly and the others gathered around him to read what he had written. The letter acknowledged that the antiwar statements the PCs had been making up to that point were not enough to help end the war. Daly remembered the rest of the letter:

> [Riate] wanted to do more than that, he was even ready to consider joining the North Vietnamese Army. . . . For a minute we all just stood there, thinking over Riate's words. Then Chenoweth and [John] Young asked Riate if they could sign the letter also.
>
> "Well, that's up to you," Riate said. "I wrote it as being only from me since I didn't know if all the rest of you had the same feelings."
>
> Then, one by one, each of us spoke out on how we felt the same and wanted to sign. My going along didn't come easily. My emotions were all for it—but it was like a double-edged knife. I think we all found it hard to believe we would even consider bearing arms against the United States.[25]

Colonel Guy, the self-appointed authority figure over the enlisted prisoners, announced the letter-signing to the prison population and threatened that the PCs could be "eliminated" or "liquidated" for their allegiance to Hanoi. The prison camp administration expressed its concern for the safety of The Eight from prisoner-on-prisoner violence through Major Do Xuan Oanh. Xuan Oanh spoke candidly with them about the open situation in Hoa Lo when prisoners were allowed to move about the facility with fewer restrictions after the Peace Accords were signed. The Eight were reassured by the awareness of the prison staff to watch out for them—and keep them safe from their fellow American prisoners.[26]

Riate's letter would later be the basis for Guy's charges against The Eight for "conspiracy against the U.S. government."[27]

JOHN YOUNG

From what we know, John Young did not come into the military with the sensitivities to race and class that Daly, Chenoweth, Rayford, and Riate did. He described himself as more middle class than the others and told Zalin Grant that he had not heard of Vietnam before coming to Fort Bragg for training. "Laos and Cambodia I knew were someplace in Asia, but I certainly could not have placed them on a blank map," he told Grant. "Nor was I familiar with the definition of communism." Young had also gone into the military unaware of the antiwar movement. "I assumed it was an isolated thing by hippies and radicals," he told Grant. His capture on January 31, 1968, was his wake-up call.

Young was an Army Specialist Green Beret captured while leading a patrol of Lao mercenaries near the Long Vei Special Forces (Green Beret) camp near the Demilitarized Zone. The camp would be overrun on February 7, handing the United States one of the most significant losses of the war. Badly wounded, he was carried by hammock for about a month along jungle trails until reaching a house belonging to the Bru minority of the Montagnard people. Young told Grant of the treatment he received from the Bru:

> They brought me breakfast and supper of soup and rice balls. I was given my own private basket of potatoes and manioc. Villagers returning from the fields always made sure the basket was full. . . . They were gentle farmers. And I began to reexamine my thoughts about the war . . . realized I knew little about Vietnam or why we were there.[28]
>
> Before this time, I couldn't have given anyone a definition of colonialism. . . . Neither could I define capitalism. I had my own definition, I guess: "The American Way of Life."[29]

By August 1968, Young had been moved to a camp at Duong Ke (called by prisoners Farnsworth, or D-1) about thirty kilometers south-southwest of Hanoi where a North Vietnamese guard gave him books to read, telling Young, "We want you to understand our side." Six of those books were authored by Americans, and Australian Wilfred Burchett's *Mekong Upstream* was particularly influential. "In the United States," he recalled, "I had heard of Southeast Asia in terms of 'the communists.' Now I was reading the human story of the people."

When Ho Chi Minh died on September 2, 1969, Young saw the

guards wearing black armbands and his chief jailer, Le Van Vuong, trying to hold back his tears. “But he couldn’t,” said Young. “He started crying. I cried too.”[30] Four days later Young volunteered to make a tape to be sent home with a visiting group of U.S. peace activists. The tape expressed his support for the antiwar movement and “marked the beginning of my protest,” he told Grant.[31]

Vuong was nicknamed “Cheese” by the prisoners because, as Young says they told him, “in America the man in charge is called the head cheese. He smiled broadly”:

> We came to like him very much. It was just that he really believed so hard, so much, in what he was doing. And I think he tried to be as fair as he could with us. I learned to look at him as a friend. He was sort of like an uncle. He taught us patience and understanding.[32]

By contrast, Cheese, as portrayed by SROs in Hubbell’s official story, was “full of hatred for his American prisoners [and] soon revealed himself to be a sadist.” Hubbell continued unremittingly in that vein, describing heinous acts of torture such as Vuong reaching into Air Force Captain Edward Leonard Jr.’s eye sockets, “fasten[ing] his small fingers on the eyeballs and squeezing them and roll them for long agonizing minutes.”[33]

Young remembered making thirty-three tapes for the North Vietnamese, and writing letters to President Nixon, to Congress, to GIs, and to people in his hometown. In letters to GIs, he explained the war and urged them to follow their own consciences.[34]

MICHAEL BRANCH AND FRED ELBERT

We don’t have the background on Michael Branch and Fred Elbert that we have for the other enlisted POWs. Young described Branch as an “extremely poor country boy” who was an Army truck driver captured near Utah Beach in Quang Tri Province and “as much opposed to the war as I was.”[35] Like the profiles of the other dissenting POWs,

Branch's class background distinguished him from the officers charging them.

Hubbell refers to a May 14, 1971, "memo signed by 'Michael P. Branch, deserter'" which read, "I've joined with a group of captured servicemen who are against the war in Vietnam." The memo advised GIs to "refuse combat or just botch up all your operations" and to "get in touch with the local people who will notify the Viet Cong [who] will get you to a liberated area [and] help you get to any country of your choosing." Hubbell calls the memo "typical" of the material read by some of The Eight while they were held at Plantation in the spring of 1971.[36]

Elbert, as remembered by chopper pilot Frank Anton, "was the strangest POW" in the Kushner group. "He had several stories about his capture," said Anton, one of which made Anton think he had been captured while AWOL. Anton also recalled an NLF radio broadcast that began: "My name is John Peter Johnson. I was in the Third Marine Division at Da Nang. I crossed over to the side of the Liberation Forces." POW Johnson denied that he was *that* Johnson of the radio broadcast, and was later revealed to really be Fred Elbert of New York.[37] Remembered by Daly, Elbert was "lying asleep much of the day. He'd work in spells, then suddenly be out of it, go off by himself, sometimes just sit daydreaming."[38]

ABEL LARRY KAVANAUGH

Tragically, Abel Larry Kavanaugh may have been the greatest difference-maker in the way the dissident POWs would be remembered in postwar America. His suicide by gun on June 27, 1973, put the public brakes on the efforts of Ted Guy to prosecute the radicals. Behind the scenes, on June 22, five days before Kavanaugh's death, top military lawyers had already submitted their final recommendation to drop Guy's charges against the PCs. But the abeyance of the dissidents' legal jeopardy only cleared the way for being recast as mental-health casualties of their years in captivity. Going forward, their images would be merged into the victim-veteran discourse already shaping public perceptions of antiwar GIs and veterans more generally. They weren't criminally "bad," but emotionally and psychologically wounded, traumatized.

Marine Sergeant Kavanaugh was captured near Phu Bai on April 24, 1968, after having been inadvertently left behind by the helicopters that lifted the rest of his unit back to their base camp. On the three-month trek to the Portholes camp (aka Bao Cao) he contracted malaria and dysentery.[39] Hubbell described Kavanaugh as a raucous prisoner in the Farnsworth camp who beat on cell walls and otherwise violated camp rules. He was placed in solitary confinement for five months. According to Hubbell, Kavanaugh emerged a religious convert, announcing

> he had been visiting with the Lord [and] knew himself to be a saint—"the thirteenth disciple." He insisted that he absolutely opposed the war, that he was opposed to all violence and the taking of life.[40]

Young remembered Kavanaugh as about five feet six, slight, very nice looking, with a swarthy complexion and raven-black hair.[41] He was a Chicano from Denver with a wife, Sandra, a daughter, and another child on the way when he died. He was Catholic with religious beliefs that Daly thought were "really fanatical."[42] Daly told Grant that Kavanaugh was "often hard to get along with," and like Riate, considered himself a "brown-skinned minority." And yet, Daly told Grant:

> When Kavanaugh talked about returning to the States, he imagined something that wasn't there when he left—a paradise. He talked about how he would be happy the rest of his life with his wife and little girl. . . . He talked about it all the time.[43]

With the option of a discharge after his repatriation, Kavanaugh had initially chosen instead to stay in the Corps. Apparently he had a change of mind. According to Daly, he was told that Kavanaugh called Riate, Jane Fonda, and Cora Weiss on the day he killed himself and said he was afraid to return to the Marines. He always thought, said Daly, that charges would be brought against the Peace Committee and, "even if they weren't true, there would be no way we would escape prison."[44]

Young told Grant that he was on his way to visit Kavanaugh in Denver when he learned that he had shot himself. Young said that he "personally felt that what [Abel] Larry did, he did for us. . . . And I knew Larry did not want to report back to the Marines. . . . But what he did was an attempt, I think, to take the pressure from us and put it on the military. He gave his life for us."[45]

Kavanaugh's suicide may or may not have been the act of selflessness that Young proclaimed it to be, but the Pentagon's dismissal of the charges against the others just six days later on July 3 suggests that his death had not been in vain. The Pentagon's press release made no mention of Kavanaugh, stating: "The dismissals were recommended because of lack of legally sufficient evidence and because of the policy of the Department of Defense against holding trials for alleged propaganda statements."[46]

However, historians Rochester and Kiley allude to a more political motivation for the dismissals. Defense attorneys, they point out, "were prepared to argue that Guy's claims were based largely on hearsay . . . and that Guy himself had been pressured into making concessions at Plantation." A trial would have spotlighted failings of the accusers, such as Guy and other SROs, who were just as guilty as those they were accusing. Dismissals of the charges, on the other hand, would "'preserve the hero image of the returnees and diffuse the radicals and peace groups who are looking for a cause.'"[47]

The concern of the Pentagon that the peace movement and dissident POWs would make common cause was, itself, a tacit recognition by the higher-ups that the antiwar expressions of The Eight had been sincere and that, though free, they were not about to recant their commitments to peace.

Daly was angry for a long time after his discharge from the Army. "It disturbed me," he wrote in his autobiography, "to see how people accepted the war as being over, while Vietnamese were still being killed every day, almost 50,000 of them by the end of 1973." "More and more," he wrote, "I found myself looking toward religion again . . . the entire experience of Vietnam, even the Communist teachings, had helped me to remember a good deal of what I'd been taught long before by the Jehovah's Witnesses, and to give me a clearer idea what to do about it."[48]

Bob Chenoweth linked up with the Indochina Peace Campaign and toured the country with Tom Hayden and Jane Fonda. Riate became a folksinger, catching the attention of Irwin Silber and Barbara Dane, two legends in the radical cultural circles of the 1960s and 1970s.[49] His song "Play Your Guitars, American Friends" was recorded on a Folkways album. After working at a Veterans Administration Vets Center, he moved to the Philippines and became an activist against the regime of Ferdinand Marcos.[50] Riate died in 1984, assassinated by some accounts.[51]

ESTABLISHMENT FEARS OF A MUTUALITY between politicized POWs and the antiwar left were understandable at the time, but two developments mitigated that likelihood. One was that the antiwar movement was feeling the burnout of ten years of hitting the barricades. Splintered by sectarian infighting with a leadership that was rebalancing its political and personal priorities, it was not looking for new fronts to forge. Moreover, and notwithstanding the Hayden and Fonda overtures to POWs, the movement had been as skeptical of the dissident POWs as were the SROs—"How do we know they were not brainwashed?"

A second development was that the psychologizing of the larger population of GI and veteran war resisters was already underway when the POWs returned, and it was proving to be a more useful means of managing their political effectiveness. Why punish behavior that can be discredited, stigmatized as a mental health problem? Why criminalize dissent that can be medicalized? The road to Post-Vietnam Syndrome's acceptance as the diagnostic category PTSD was already being paved, so why not put the POWs on it along with other

Vietnam veterans? The Kavanaugh suicide, fortuitous for the military and political elites wanting to see the war in a rearview mirror, shifted the paradigm from "bad" to "mad," from villain to victim, and displaced from social memory the image of warriors, including some POWs, who turned against their war.

The history that POWs conscientiously opposed the war before and after their return from Hanoi was not so much erased or suppressed from public memory as it was displaced or overwritten by other, related story lines. The pathologizing of their dissent as a symptom of trauma was the death blow to the sense that POW dissent was conscientious. However, the psychologizing of their protest was itself a refinement of the idea that in-service, veteran, and POW resisters to the wars they were sent to fight had been misled by foreign propaganda, even "brainwashed" by the enemy. It was an idea hatched in the years after the war in Korea to explain why some U.S. POWs made statements denouncing the war, and even considered staying, or opted to stay, with their captors rather than come home when they could.

In the next two chapters we explore the trajectory of the brainwashing thesis arising out of American Cold War obsessions with communist mind-control and internal subversion, and its extension into the years of the war in Vietnam and the period that followed, during which the social memory of POW dissent would be written into obscurity.

4

The Manchurian Candidate Stalks the Homeland: Hollywood Scripts the POW Narrative

Whether cowards, opportunists, or true believers . . . the "PCs" had become totally controlled by the enemy.
—Rochester and Kiley, *Honor Bound*[1]

History books, news media, memoirs, novels, film, and folklore all play roles in shaping American memory of the war in Vietnam. But few of these play more powerfully than Hollywood films and, within that medium, the images of POWs are particularly intriguing.

During the months and years of their lockup, Americans had little news about the prisoners. The trickle of letters and photographs that reached the United States raised suspicions about their verity. Who *really* wrote the letters? Were there secret messages encoded in the photographs? With nothing but their imaginations to fill in the blanks of time and place generated by the captivity of their flyers and fighters, Americans turned the Vietnam POWs into fantasy figures cast in the myths and legends passed down from previous wars. And no war played more prominently in the imagination than the one in Korea, 1950–1953. The brainwashed Korean War POW—a weak-willed traitor, a turncoat defector, a veteran *cum* mind-controlled spy and saboteur—was the archetype that could disparage antiwar POWs and veterans of the war in Vietnam, while at the same time fill theater cashboxes.

Historians Stuart Rochester and Fredrick Kiley wrote *Honor Bound* in 1998, well after the notion of brainwashing, that is, "mind control," had lost legitimacy in professional psychological literature.[2] They nevertheless dared to write what they did because associating the captives held in Asia with brainwashing was so inscribed in American culture as to be a kind of common sense, and anyone questioning it was met with ridicule. Indeed, if Rochester and Kiley had written the opposite—that dissenting POWs had acted out of principle—it is doubtful that the Naval Institute Press would have published their book.[3]

Brainwashed POWs was the paradigm, the background assumption, used by Americans for making sense out of the fragments of information they had about what went on behind the bars and walls in Hanoi. Brainwashing was the centerpiece of the Korean War backstory that U.S. pilots had taken into their Vietnam War experience: U.S. Korean War prisoners had "caved" to the persuasiveness of their captors; their confessions of war crimes and renunciations of loyalty to the homeland had *not* been coerced. Stigmatized in the press and film as collaborators, Korean War prisoners had been portrayed as men-of-weak-character whose shortcomings were derived from a permissive and hedonistic society. As an antidote for that slur, Hollywood went overboard after the defeat in Vietnam to popularize the self-aggrandizing images that senior officers made upon their release from Hanoi in 1973—*they* had distinguished themselves as real-deal tough guys who had stayed at war even when tortured.

American amnesia about the POW experience in Vietnam is a product of popular film fare and critical film studies that has been abetted by overlooking that there were divisions within the POW group, and not calling out the silence of filmmakers on class issues that bred the animosity separating the prisoners. The films through which Americans would know and remember the POW experience of the war in Vietnam were largely based on the memories of a few SROs and the conveyance of those memories, through interviews with historians and journalists, onto the desks of screenwriters.

THE MANCHURIAN CANDIDATE

The filmic narrative with torture at its core is itself not unproblematic given that it is derived from the memories of elite officers, going back to the years 1965–69 when the tortures reportedly occurred. Moreover, we know from studies of memory that the events remembered are also constructs of the expectations that the memoirist brings to the event. In the case of the Hanoi captives, those expectations were set by what the prisoners knew, or thought they knew, about what happened to prisoners in Korea, much of which, in turn, had come to them through film.

Robinson Risner tells in his memoir *The Passing of the Night* that he entered the prison experience confident he could take "anything they could dish out." He wrote, "They could torture me to death, and I would never say anything." His self-certainty was based on his Air Force survival school training and reading "fiction magazines where people had been tortured into unconsciousness and never uttered a peep. I believed without question I was as strong as they were and that I could take it."[4]

Risner doesn't say what "fiction" he had read or what the survival school curriculum included, but it is not unlikely that the lesson plans were laced with some literary enhancements. "Dubious wisdom about American Korean War POWs," concluded Craig Howes, "profoundly influenced military policy, and eventually the way many Vietnam POWs viewed their own captivity." Howes went on to cite the 1962 film *The Manchurian Candidate* as "contain[ing] the essence of received [public] opinion" that "skilled communist indoctrinators had brainwashed American soldiers into renouncing their country."

The Manchurian Candidate is based on a 1959 novel by Richard Condon in which Staff Sergeant Raymond Shaw (Laurence Harvey), captured in Korea, has been programmed by the Chinese to carry out an assassination plot in the United States after he returns home from the war. Shaw's domineering mother (Angela Lansbury) is part of the plot, and he is portrayed as too weak—even incestuously so—to resist her entreaties. Shaw's target is his mother's new husband, who is running for president. Major Bennett Marco (Frank Sinatra), who had been captured with Shaw and is unsettled by his own nightmares about the war experience, perceives the conspiracy and races to the Madison Square Garden campaign rally to stop Shaw.[5]

THE KOREAN WAR POW STORY: WRITTEN IN HOLLYWOOD OR VIENNA?

The received public opinion on Korean War POWs that inspired *The Manchurian Candidate* had its own antecedents in film fare.[6] In *The Rack* (1956), Paul Newman played Army Captain Ed Hall Jr., who came home from Korea after being held prisoner for two years. His brother Pete had been killed in the war and his father, an Army colonel played by Walter Pidgeon, is devastated to learn that Ed will be court-martialed for giving "aid and comfort to the enemy under no duress or coercion." Under questioning by his own attorney, Hall confesses to having made propaganda statements demanded by the Communists and acknowledges he suffered no physical torture prior to doing so. With his conviction seemingly assured, Hall's defense attorney then prompts Hall to recall the prison interrogations that had elicited details of the childhood anxieties that were caused by his father's coldness.

Hall: [The Communists] brought me some paper and a candle and said I should write—but only about myself. They also questioned me. They kept me in the cellar. It was freezing. They said I had to sign a leaflet right then or they'd leave me alone for the rest of my life.... They used to come to the door and asked how I liked being alone. I'd bite my hand to keep from answering. And, at night, they came and said there was a letter for me. At first, I didn't believe them because nobody wrote to me before . . . and they said it was because nobody cared for me.

Attorney: What was in that letter?

Hall: It was very short. It was from my father. He asked me why I hadn't written to him lately.

Attorney: Is that all?

Hall: No. He said my brother Pete had been killed. And at that moment I knew I couldn't hold out any more. So, I said okay, yes, I'll sign anything . . . but I gotta get some sleep and I can't be alone anymore.

Attorney: So, in other words you collapsed.

Hall: Yes.

Attorney: Now, Captain, when they had told you to write, was that in the nature of an autobiography?

Hall: Yes.

Attorney: And what was the purpose of all this? Do you know?

Hall: Trying to find out about me.

Attorney: Trying to find out something they could use? Trying to find out some special thing they could use to break you. And finally, you gave it to them, gave them the key they were looking for.

Hall: Yes.

Attorney: (Reaches for a document) Is this not a true copy of your autobiography you wrote, what you wrote after being held in miserable conditions, seeing the letter from your father, and telling them what you did in your autobiography?

Hall: Where did you get that?!

Attorney: I've marked here a passage I'd like for you to read to the court, read out loud.

Hall: No, I can't.

Judge: The witness is instructed to read.

Hall: (Reading) When I was a kid, I remember our home—

Attorney: (Interrupting Hall) We can't hear you. Your hand is covering your mouth.

Hall: (Continues to read) . . . was a place where we had to whisper, where we shut the door. My mother was sick so much she couldn't be with us. It was a lonesome time for Peter and me. We kept waiting for our mother to get well so she could be with us. Then she died. My father was away in the Army most of the time. He left us with a bulletin board with reminders on it and our housekeeper gave us a star every day if we were good and our father looked at them when he came home. Our father taught us a lot of things soldiers should know—how to keep our room neat, how to take orders, and that soldiers don't complain. As far as I remember (choking up) my father never kissed me or held us (breaking down, sobbing, unable to finish reading).

Attorney: (Picking up the document and continuing to read from it) . . . or held us or doing the other things we would see fathers doing. I never felt warmth. I'm as strict with myself as he was. I can't seem to love anybody. I love Mother but she is gone. I love my brother but he's not here. I wish my father had given me the chance to show him how much I loved him. If this loneliness can kill me, I hope it does now.

Judge: Does the defense wish to rest?

Attorney: No, your honor.

After a court recess, the prosecutor cross-examines Hall in an exchange that would lay the template for the weak-character explanation for POW collaboration and the deficiencies in American society behind it.

Prosecutor: The thing that made you collapse was loneliness—isn't that what you said?

Hall: Yes.

Prosecutor: Now this matter of loneliness interests me. I'd like to examine it a bit. Now, Captain Hall, can you tell me about some special time when the loneliness was especially bad, sometime before you went to Korea?

Hall: What do you mean? A special day or what?

Prosecutor: Yes, a special day would be fine.

Hall: Well, some days were worse than others, I guess.

Prosecutor: You mean there was no one day when the loneliness was just so bad you couldn't stand it? (Shouting) One day! (Softly) What was loneliest day of your life?
Hall: The day my mother died.
Prosecutor: The day your mother died. The autobiography you wrote—the Chinese found your weakness, a very lonely boy—they beat you with it.

"A very lonely boy"—the "boy" puts the prosecutor's words in relief. Hall was a decorated soldier for his service in the Second World War and had suffered a mortar wound in Korea. "Boy," uttered here, as the camera panned the "oh-that-poor-boy" expressions on the faces in the courtroom and focused on Hall's own childlike sniveling and looks of shame, was clearly meant to bring "mama's boy" to mind for the viewers. That same "boy" was called forth in 1973 when Mike Wallace admonished Gene Wilber on CBS's *60 Minutes* with "You weren't a child."

The Rack's contribution to the theme of young male personalities made vulnerable to authoritarian figures might appear, a half-century later, as remarkably creative, a manifestation of a screenwriter's overactive imagination. In the 1950s, though, it was in step with the efforts of social scientists to understand what had lent whole populations in Europe to follow Hitlerian leaders. The 1950 book, *The Authoritarian Personality*, written by Theodor Adorno and others associated with the Frankfurt School of Critical Theory, argued that the domineering personalities commonly associated with Fascists had their dialectical Other in the submissive personalities that fell in the thrall of dictators. Informed by Freudian psychoanalysis, the book gave special attention to the institution of family, the site at which oedipal conflicts forged the personal psychologies that carried into adult life.[7]

Viewers of *The Rack* got early notice that the filmmakers were headed in this direction when the prosecutor called to the stand a witness who had been imprisoned with Hall. The witness testifies that Hall had signed a compromising statement for the guards. Under cross-examination, the witness says Hall seemed to have changed after having been separated from the others for the period when Hall was in solitary confinement. The attorney pressed the witness on the changes he observed in Hall.

Attorney: Now, did you notice any great difference in Captain Hall? Had he changed in that period?
Witness: Well, yes. He seemed terrified.
Attorney: Had you ever seen anyone as terrified to the same extent?
Witness: Some years ago, when I was very young.
Attorney: Where?
Witness: Dachau.
Attorney: Dachau. The death camp in Germany!
Courtroom: [Gasp!]

When the prosecutor objected to the implied analogy between fascist Germany and the American culture responsible for Hall's collaboration, the judge ordered it stricken from the record. But the defense attorney returned to the point in his closing argument:

Attorney: (To the Jury) Gentlemen, if there is guilt, where does it lie? In the small defect under pressure as with Captain Hall? Or do we share it? At least those of us who created a part of a generation which may collapse because we have left it uninspired, uninformed, and like Captain Hall, unprepared to go the limit. And now we must judge Captain Hall.

BEYOND *THE RACK*

It would be almost twenty years before Hollywood would begin portraying POWs from the war in Vietnam. But the die was cast in *The Rack*, which displaced to prologue the story of what happened in the POW camp, and foregrounded the story of the repatriated captive coming home to a nation skittish about the incipient corruption of post–Second World War affluence and the Cold War climate in which McCarthyite zealots looked for communists behind every door. The narrative of homefront flaccidity carried forward through *The Manchurian Candidate* and into the popular culture that would forge the mindset of military men headed for Vietnam. Thence, it is the narrative through which the American public would remember that the war in Vietnam was lost to political liberalism, permissive parenting, economic entitlements, and a pernicious feminism that sapped the national will-to-war. It was a war-at-home with the front-line running through the psyches of the captives in Hanoi just as it did through Captain Hall's.

Fifty years later, *The Rack* might be retained on lists of a few film buffs but otherwise it is forgotten. However, its all-star cast dispels doubt that it reached into the heart of America's postwar reckoning with Korea. Walter Pidgeon had won Best Actor Oscar nominations for *Mrs. Miniver* (1942) and *Madame Curie* (1943). Lee Marvin as the Army captain who testified against Hall at the trial was already an established figure in war movies and recognized for his 1953 casting with Marlon Brando in *The Wild One*. Paul Newman was just two years away from his breakout role opposite Elizabeth Taylor in *Cat on a Hot Tin Roof.*

Similar to *The Rack, Time Limit* (1957) depicted Major Harry Cargill in a cold and forlorn POW camp, brainwashed by the communists into delivering Marxist lectures to his imprisoned buddies. Speaking in a rote and mechanical fashion, Cargill's addresses signify the mind-snatching power of communist indoctrination that was the hallmark of the POW film genre well into the 1980s. As with *The Rack,* the prison camp scene is only a flashback setting for the drama of Cargill's postwar trial for collaboration, the film's main story. In the end, Cargill is relieved of charges when he is shown to have cooperated with the communists as a strategy to save his fellow POWs from execution.

Time Limit also had the star power that ensured it a wide viewership. As its protagonist Major Cargill, Richard Basehart had already played in many war movies, including the 1951 *Fixed Bayonets*. For Cargill, he was awarded a BAFTA, the British Oscar. June Lockhart, playing Mrs. Cargill, had won a 1948 Tony and was embarking on her role as Timmy Martin's mother in the *Lassie* television series. Richard Widmark as the colonel who interrogates Cargill for collaboration was a 1947 Oscar nominee, and played opposite Marilyn Monroe in the 1952 *Don't Bother to Knock.*

In the chronology of influence, the Korean War films are important first for the influence they had on the mindset that U.S. fighters took with them to Vietnam. Since many were veterans of the Korean era, it's hard to imagine they did not watch with great interest the way Hollywood represented the war and the men who fought it; the star-studded cast of a film like *The Rack* would have added to its appeal.[8] Second, Korean War film fare would influence the American film-makers who turned out the first round of films set in the years of the war in Vietnam. Third, the films' cultural framings shaped the way Americans thought about war in the post–Second World War era, which in turn provided the cues that looped back to the screenwriters looking for story lines fitting for Vietnam and the expectations of the troops training for Vietnam.

THROUGH FRENCH INDOCHINA: THE CIRCUITOUS SEGUE TO HOLLYWOOD

Most of the story lines that would characterize the Vietnam POW films were developed in the Korean War films of the 1950s. The transition from one to the other, however, was eased by some films linking story lines across temporal and spatial boundaries. *Dragonfly Squadron* (1954), for example, was set in Korea but glanced to French Indochina for riffs that may have jump-started the imaginations of pilots, filmmakers, and theatergoers of what awaited them in Vietnam: torture and spousal fidelity. Chuck Connors, already a two-sport legend for having played for both the Brooklyn Dodgers and the Boston Celtics, plays an Army captain whose unit comes to the rescue of American aid workers about to be overrun by the Chinese communists. The group's leader, Dr. Stephen Cottrell, had been previously captured in French Indochina and held prisoner by the Viet

Minh; as torture, the communists there had jammed spikes under his fingernails and ruined his hands for surgery. Mistakenly reported as killed in Indochina, Cottrell is now in Korea where he unexpectedly encounters his wife, who is there with the medical aid group and romantically involved with Air Force Major Matt Brady, assigned to train Korean pilots.

The primitive savagery attributed to the Viet Minh in *Dragonfly Squadron* was game-changing because earlier films that portrayed German POW camps had depicted more passive deprivations, such as inadequate food. The physicality of those representations of camp life in Indochina may have been metaphoric for the materiality of the Second World War's ground war versus the ideological nature of the Cold War—poignantly captured in the characterization of Vietnam as "a war for hearts and minds." POW films of the Second World War era such as *Stalag 17*, moreover, presented German prison administrators as hapless bunglers presumably too dumb to know about the mind-body chemistry of physiological torment.[9] The bridge to the psychologizing of torture, crossed as it was, with the Orientalizing of its agents, also had the effect of racializing the conflict between SROs and the dissident POWs, casting the latter as race traitors as well as military turncoats.[10]

Dragonfly Squadron also may have introduced the theme of disloyal POW wives back home. Second World War films did not give wives the prominence they would get in Vietnam war films, perhaps because the two to three years of captivity in Europe and Japan was a comparatively short time compared with the as much as eight years for the first pilots shot down in Vietnam.[11]

Another movie was *The Lost Command* (1966), set in Indochina, not Korea, just after the French defeat at Dien Bien Phu in 1954, which cued the U.S. intervention in the region. It's an important film because it introduces three more story lines that will typify what will come out of Hollywood's rendition of the Vietnam POW experience: dissension within the officer ranks; the role of class background; and the lure of the racial or ethnic Other as subtext in the issues of collaboration with the enemy.

In *The Lost Command*, Lieutenant Colonel Raspeguy leads a French regiment in a last-ditch stand at Dien Bien Phu. Raspeguy is frustrated that he has not received reinforcements soon enough, forcing him to lead his men in retreat. After capture by the Viet Minh, the unit is led on a forced march before being released when the French withdrawal from Indochina is negotiated. On his return to France, Raspeguy's troop ship stops over in Algeria where he learns that the unit has been disbanded and he has been relieved of his command. Contemplating his post-military life, Raspeguy returns to the peasant village in the French mountains where he grew up. In conversation, he acknowledges the social distance between his rural peasant roots herding sheep and the French army officers he was expected to follow.

Later, Raspeguy gains an audience with a government minister to plead his case for reassignment. Entering the office, he declines the minister's invitation to sit down saying, "He learned to stand as a shepherd boy." His eyebrow raised by the remark, the minister reads from a report that the Colonel has the respect in the field of his junior officers but that he had disobeyed orders at Dien Bien Phu. "It's your superior officers with whom you can't get along, and that has to stop," the minister commands.[12] The minister offers Raspeguy a new post, leading a unit in Algeria fighting the insurgent liberation forces. Engaged in what would become famous as the Battle of Algiers, Raspeguy faces a young Algerian fighter who had fought with Raspeguy's unit in Indochina. From there, the film revolves around the ethical dilemmas posed by personal, ethnic, and national loyalties.[13]

THE FIRST AMERICAN POW FILMS SET IN VIETNAM

It is no surprise that historians interested in what happened to the American memory of dissent within the Vietnam POW population would assume that movies had something to do with it. No surprise either that their study of Hollywood influence in that important piece of history would begin with the 1987 film *Hanoi Hilton*, thinking that it was seminal in the POW-themed genre.

In fact, *Hanoi Hilton* came out fourteen years after the POWs returned and even more years after films made in 1971 first referred to POWs in Vietnam. Most of those early films dispensed with the war itself and the history of the POW experience and featured the coming-home travails of the now ex-POWs.

In *The Forgotten Man* (1971), Dennis Weaver plays Marine Lieutenant Joe Hardy, who comes home after five years as a POW to find his wife remarried and his seven-year-old daughter adopted by her new husband. The early scenes are filled with gut-wrenching father-daughter efforts to resolve the quandaries they face. The home-front betrayal encoded in his wife's infidelity fuses with Hardy's doubt that the government had tried its best to rescue the prisoners in Hanoi. Those specters of infidelity set off his paranoid responses. Now facing government authorities trying to take his daughter from him, Hardy returns to combat instincts, kidnaps his daughter, and is killed in a shootout with the police.

Welcome Home Johnny Bristol, another 1971 film, opens with Johnny Bristol held in a bamboo cage. Bristol is remembering Charles, Vermont, the idyllic small town where he grew up. Upon repatriation, Johnny is hospitalized at a Boston VA hospital where he woos his nurse, Ann, with his descriptions of Charles. As the story unfolds, however, we see that there is no Charles, Vermont, and never was. Johnny had fantasized Charles as the mythical white-picket-fence America to which he hoped to return.

Strikingly, these two early Vietnam POW films stay on the road paved by their Korean War precursors, *Time Limit* and *The Rack*, in framing their content as a coming-home story and reducing the war to background. Both were made for television, thus assuring a wider viewing than they may have received in theaters, and, in turn, increasing their influence on the just emerging postwar representations of the war-at-home narrative that will dominate the American memory of the war years. *The Forgotten Man* picks up the spousal-loyalty thread left by *The Dragonfly Squadron*, and *Welcome Home Johnny Bristol* writes the subtext, albeit through Johnny's war-borne psychosis, that the Norman Rockwell America from which GIs departed for Vietnam was rotting, due to its discard of traditional values.

Although these two films were released two years before the wholesale repatriation of the 591 Hanoi prisoners—too early for them to have touched on the story of dissent within the prison population—they ignored completely the figures of GIs who had been captured in the South years earlier and returned home opposed to the war. These early films could have featured, for example, ex-POW George Smith, who was released in 1965 after two years of captivity in the South and wrote *P.O.W.: Two Years with the Viet Cong* about having come to consciousness about the war during captivity. Or they could have told the stories of James Jackson, Dan Pitzer, and Edward Johnson, who were freed by antiwar activist Tom Hayden in 1967 only to return home to face charges that they had been brainwashed.[14] Rather than presage the fact that the Peace Committee, the hard core of the Hoa Lo dissenters, was composed of GIs captured in the South and forge a new template for the POW narrative, filmmakers ignored the story.[15]

POWs, THE NEW LEFT, AND THE BURDEN OF BRAINWASH

The exclusion of the dissident POWs from Hollywood story lines was not a given. After all, antiwar veterans who had not been POWs would soon be marqueed in the 1978 *Coming Home*, starring Jon Voight. His character Luke comes out against the war while recovering from wounds in a VA hospital. But *Coming Home* was an exception, the explanation for which illuminates the complexities surrounding the cultural representations of the POW dissenters.[16]

Coming Home was the project of actress Jane Fonda, who had gotten involved with uniformed war resisters and antiwar veterans while living in Paris in the mid-1960s. When she returned to the States for good in 1969, she became an ardent supporter of the GI Coffeehouse Movement and helped fund the 1971 Winter Soldier Hearings into U.S. war crimes. In short, Fonda's star power and financial resources gave Vietnam Veterans Against the War (VVAW) the media visibility and celebrity imprimatur that magnetized its reputation and drew thousands more Americans to the cause of ending the war. Antiwar veterans were a hot ticket for the movement and the box office—unless they had spent their Vietnam years in Hoa Lo prison.

The antiwar POWs were not the darlings of the antiwar movement that the whistleblowers of the Winter Soldier Hearings had become. In the first place, most were pilots who had rained hell on the Vietnamese. And second, most were old enough and well enough educated to have known better than to have committed the airborne atrocities they did—or they should have known better. The antiwar statements made by dissident pilots, moreover, came *after* their bombing raids left civilians terrified and North Vietnam's urban infrastructure shredded, a fact that hardly burnished their sincerity. Most of the dissident POWs were not pilots, of course, but the pilots were clearly the stars of the coming-home show, making "guilt by association" (with them) a kind of collateral jeopardy borne by the others: *all* these guys are professional killers undeserving of sympathy or solidarity went a line of left-wing thinking.[17]

Additionally, war opponents were as inhibited by Cold War fantasies of communist brainwashing as anyone else. Had some captured pilots really come out against the war? Or had their statements been coerced? Had they been brainwashed? And would a peace activist quoting a recalcitrant POW be dismissed as a communist dupe—brainwashed? Would a left-wing acceptance into the movement of the antiwar POWs be spun by the Right as proof that the movement was a front for International Communism?[18]

This is the kind of Cold War baggage carried by peace travelers to Hanoi. Their meetings in Hanoi with captive pilots would be seminal in forming each group's impressions of the others.[19]

PEACE TRAVELERS MEET THE POWs

Carol McEldowney and Vivian Rothstein kept journals on their 1967 trip to Vietnam. McEldowney had been active in SDS at the University of Michigan; she was present at the Port Huron retreat where the SDS manifesto, known as the *Port Huron Statement*, was written. In 1964, she moved to Cleveland to work with the SDS community-organizing project, ERAP (Economic Research and Action Project). Rothstein had gone to UC Berkeley and been involved in the Free Speech Movement. After civil rights organizing in Jackson, Mississippi, in summer of 1965, she joined the ERAP campaign in Chicago.[20]

In 1967, McEldowney and Rothstein were invited by SDS leader Tom Hayden to attend a conference in Bratislava, Czechoslovakia, that brought U.S. antiwar activists together with representatives of the North Vietnamese government and the National Liberation Front (NLF). As the conference ended, the Vietnamese invited Hayden and six other Americans to visit North Vietnam. McEldowney and Rothstein were among the six.[21]

McEldowney's journal is speckled with signs that "brainwashing" weighed heavily on the minds of the Hayden delegation. On her fifth day in Hanoi she wondered if the cultural panic surrounding communism in the United States would taint whatever her delegation reported upon return to the States: "For me (for us?) the problem will be to learn to communicate what we've learned to people in the U.S. without seeming brainwashed."[22]

The contemplation of self-censorship lest their testimonies be dismissed as biased was difficult, but the specter of brainwashing presented the travelers with a still more pernicious face. The New Left, of which SDS was the central component, and to which the Hayden delegation was fundamentally committed, had itself been born out of dissatisfaction with communists who had dominated its Old Left predecessor, the Communist Party of the United States, the very communists publicly reviled for simple-minded servility to the Soviet Union and the practice of mind control within its organization. With those strains of Cold War anti-communism coursing through its own veins, the Hayden delegation faced a dilemma: how to keep their guard up against expected attempts by their communist hosts to brainwash *them* while simultaneously processing their experiences for presentation at home in a way that challenged public paranoia about communist duplicity and mind control. "All of us must avoid doing what the pro-Soviet people [in the CPUSA] did in the 1930s," McEldowney wrote in her journal, before adding that their public presentations would have to acknowledge their "awareness of some the restrictions and less attractive characteristics [of communism]."

The Americans went into their meeting with POWs suspicious of both their Vietnamese hosts and the prisoners. Would the prison administration try to prettify the prison conditions with testimonies coerced from the prisoners? Were the prisoners handpicked for interviews because they had been brainwashed or terrified into mouthing the communists' line? "We had a lot of discussions about whether the pilots were truthful—were they really treated well or were they being forced to pretend?" Little did they know that the prisoners were suspicious of them, lest they had been brainwashed by the communists. Sailor Doug Hegdahl, who had been captured in May, asked the delegation several times, wrote McEldowney, "who sent us, who financed us, and whether we were communists."

In the end, the Cold War infection of their own political perspectives may have been the insurmountable barrier to leftist affinity with the antiwar POWs.

The Bratislava group met Air Force pilot Larry Carrigan on October 12, 1967; he had been captured just a month before meeting the group. In her journal, Rothstein wrote:

> [Carrigan] asked us why we [are] against the war—we said because Saigon gov. is not what VN people want, that it is brutal & we have no right to be here.
>
> He said he agreed. That he really knew very little about VN before he came here.
>
> Said we should go home & not be belligerent but try to talk to the Amer. people & explain abt. the Geneva Accords.

"After we spoke to pilots," wrote Rothstein, "Tom [Hayden] tried to explain that it was very 'disturbing' because we didn't know what to believe of what the pilots said or how to use it as propaganda back home." She continued: "Tom didn't say exactly what he meant—that he was sure the prisoners were treated well & that they are not getting brainwashed but that we think pilots who are as anti-war now as Carrigan are talking like that because they've been taught to by the Air Force once captured that they are not politically more aware."

Hayden's twist on the POWs' veracity is eye-catching, but McEldowney's record of the Carrigan interview minces fewer words:[23]

> After a while, I didn't believe anything he said (and I withdrew). . . . I really hated him—saw him as the white guy from the Southwest who would drop bombs on negroes in a race riot.[24]

A few lines later, she wrote, "I felt for the first time that we were being *used* by [the communists]."

Following the meeting, Rothstein summarized her feelings:

> The experience was rather awful. We all fell into a closeness with the pilots on the basis of being Americans—we forgot where we were and who we were. Not until the middle of Carrigan's talk did I realize not to trust what he said at all. I felt a little like I was in the *Manchurian Candidate*—with the oriental communists around us and us in the situation of being brainwashed. It was caused by all the stereotypes I've ever known [and heard] about Chinese and Korean communists and their brainwashing.

McEldowney also inserted notes from John "Jock" Brown's diary in her journal. Brown was a minister who had joined the group in Czechoslovakia. Amid rambling notes about Carrigan's views about antiwar demonstrations—he said they were "great" but questioned their effectiveness—and the need to educate Americans about the Geneva Accords, Brown recorded the following:

Question for Carrigan from Vivian (Rothstein): "Will they say you've been brainwashed?"

Carrigan: "I don't know what the military will say, but do I seem brainwashed (laugh)? I was never brainwashed."

LEFT BEHIND: REBEL POWS ON THE CUTTING-ROOM FLOOR

With no antiwar Left to nurture the political capital of dissenting POWs who returned home, and no Jane Fonda to champion them, the POW story would be hijacked by the political Right to stir public unsettledness about "what really happened over there." The Right's agenda to feed suspicions that the government was not telling the truth about the war as they cast aspersions on the loyalty of the antiwar movement was prelude to the revanchism of the Ronald Reagan presidency that would popularize the paranoia about "Washington insiders" as seditious fifth columnists. Hollywood's contribution to that agenda would be twofold: first, displace the image of the dissenting POWs, and eventually all real-life POWs, with that of mythical POWs abandoned by the government and left behind in Southeast Asia; and second, recast the image of radical veterans from the political problem they were into a mental health problem.[25] Wilber, Chenoweth, and their brothers-against-the-war who had formed Vietnam Veterans Against the War years earlier, were better understood as traumatized victims than as warriors transformed into peacemakers by their experience in Vietnam.

Hollywood's disinterest in the story of peacemaking POWs is understandable. The Cold War hysteria about brainwashing that was ginned in Korean War films like *The Rack* and *The Manchurian Candidate* was now blowback that inhibited artistic and marketing judgments about what the theater-going public would pay for. Intimations that not all the POWs had come home dampened interest in the accused collaborators and even their hardcore accusers. The POW *du jour* was the one unacknowledged by the government and forgotten by the public, the one who did not come home.

A spate of films made after 1973 featured the coming-home travails of the now ex-POWs: *Mr. Majestyk* (1973), *Rolling Thunder* (1977),

Ruckus (1980), and *Some Kind of Hero* (1981), all following that pattern. That it took four years before the first explicitly POW-themed film would be made, in 1977, suggests that Hollywood needed time to digest the loss of the war and decide what to do with the POW story.[26]

Good Guys Wear Black in 1977 made a statement: the government had abandoned POWs in Southeast Asia; the war was not over so we—veterans, now citizen-soldiers—would have to go back, rescue the POWs, and finish the job. The raiders, arriving at a bamboo-prison compound emptied of prisoners—but occupied by Viet Cong ambushers—have been set up, a betrayal blamed on Washington by the unit's leader. Metaphorically, this stab-in-the-back given to the mission contributes to the betrayal narrative for the loss of the war then building in the country, while the motif of duplicity that enveloped it primed a public inclination to view POW dissenters as traitors.

Uncommon Valor (1983) followed suit. A television news report in the background tells listeners that 2,500 POWs are still unaccounted for and "until they're all back . . . the war will never be over." Washington is doing nothing; Col. Jason Rhodes (Gene Hackman) thinks his son Frank is alive among the missing; with the backing of a rich donor, he assembles a unit of Vietnam veterans to go searching for Frank. Trekking into Laos, the unit arrives in a decimated village where it finds only the skeletons of its inhabitants said to have been gassed. Further on, a POW camp is raided but Frank is not among those brought home.[27]

As a genre, the POW/MIA films consolidated Nietzschean themes flowing from 1950s films such as *The Rack* that called out the effeminized post–Second World War culture that had sapped American will-to-war. Filmic images of POWs languishing in bamboo cages were metaphors for the American greatness and masculinity lost in Vietnam that awaited recovery and return. Ronald Reagan's proclamation that "government is the problem, not the solution" was a call to arms for avenging Ramboesque vigilantes who could counterweight the tax-paying public's reluctance to fight again: the "Vietnam Syndrome." Reagan's signature slogan—"It's morning in America"—summoned the back-to-the-future sentiments seeking restoration of a prelapsarian America whose men had been waylaid by the women's movement, whose work ethic had been eroded by government entitlements, and whose pride in the military victories of the Second World War had been insulted by the antiwar movement and the Vietnam veterans associated with it.[28]

THE POW-RESCUE AND MIA-recovery films were not a departure from the brainwashing story line of *Manchurian Candidate* fare that dated back to the post–Korean War years. Rather, the screenwriters took the psychological premises of brainwashing beyond the colloquial expression of "psychologizing the political," popular in the early 1970s, and created a new diagnostic category initially known as "post–Vietnam syndrome" and later as Post Traumatic Stress Disorder (PTSD). Cast in film as captives to be found and rescued, POWs were rendered as passive victims or even hapless losers who needed saving by fantasy figures cut by Sylvester Stallone's Rambo and Chuck Norris's Colonel Braddock.

Written out of Hollywood scripts and, going forward, out of public memory, representations of dissident POWs would be dissolved into the larger set of antiwar veterans that cultural workers, journalists, and psychiatrists were busy pathologizing as mental and emotional casualties of war who were more deserving of sympathy than solidarity.

5

Damaged, Duped, and Left Behind: Displacing POW Dissent

By the end of the 1970s, Vietnam veterans were commonly portrayed in film and news reports as casualties of the war, their military mission sold out on the home front and their homecoming marked by ingratitude and condemnation. Representations of POWs followed a similar path. The hero-prisoner imaging that had dominated the news through Operation Homecoming soon faded as a war-fatigued public turned to affairs it had put aside for the draft, service, and protests against the war.

What interest there was in POW dissenters merged into the mental health discourse that was medicalizing GI and veteran dissent as symptomatic. It was trauma, not politics and conscience, that moved in-service resisters. If there was energy for the war after 1973, it was mustered by conspiratorial rightists around the myth that some POWs had been left behind as Washington exited Vietnam.

MEDICALIZING DISSENT

The template for the representation of POW dissidents had been forged by news organizations and mental health professionals in their coverage and treatment of the veterans who had already come out against the war by the time POWs arrived home.

In-service resistance had been rife since the war's earliest years. That resistance was expressed through claims to conscientious objection, refusals to deploy, collaboration with civilian peace activists at off-base coffeehouses, and efforts to organize opposition through the GI Press, a network of antiwar newspapers. In Vietnam, war resisters displayed antiwar symbols on clothing, sabotage of equipment, refusals to carry out orders, and acts of violence against superiors, known as fragging.[1]

Unsurprisingly, authorities in and outside the military sought to prevent, disrupt, and punish acts of dissension. GI coffeehouses were declared off-limits and raided by local police, radical newspapers were confiscated, peace symbols were banned, their wearing punished as Article 15 violations, and court-martial charges were brought against the most serious offenders. Predictably, attempts to suppress dissent bred more misbehavior, and by the last years of the war the low level of troop discipline threatened military operations.[2]

Officially, the military denied that it had a problem. The investigative reports it commissioned on the problem in 1970 and 1971 were not made available until after the war. Journalists were dispatched to Vietnam to find and report *war* stories, not *antiwar* stories. Until veterans returned with eyewitness accounts of breakdowns in unit discipline that might be affecting operations, news organizations either remained oblivious to the emerging rebellion or ignored it.[3]

But the news came out. In early May 1969, the news service UPI carried a lengthy report on the GI antiwar movement that included photographs of coffeehouse scenes and stories from underground GI newspapers. On May 23, *Life* featured a story about what it called "a widespread new phenomenon in the ranks of the military: *public dissent*." In August, the New York *Daily News* reported the refusal of an infantry unit in Vietnam to continue fighting. On November 9, the *New York Times* ran a full-page advertisement that was signed by 1,365 GIs opposing the war and included the rank and duty station of each signer.[4]

That the American people knew all of this as it happened underscores several questions: What happened between then and now? What happened to the awareness and memory of such widespread resistance to the war within the military?

The answers to those questions help explain the expurgation of POW resisters from public memory. In the closing years of the war, establishment leaders began to worry about the legacy left by a generation of warriors who turned against their war. In short form, the Nixon administration, news organizations, and leading cultural institutions—Hollywood filmmakers in particular—endeavored to cast shade on the authenticity and reliability of their dissent.

OSTRACIZING DISSENT

The April 15, 1967, mobilization against the war, known as "Spring Mobe," had brought together six veterans to form Vietnam Veterans Against the War. The VVAW presence six months later at the October March on the Pentagon and support for the 1968 presidential campaign of antiwar candidate Eugene McCarthy established the group as a powerful voice for peace in Vietnam. Unable to directly suppress and punish dissenting views, the Nixon White House and its pro-war minions sought instead to drive a wedge between the radical veterans and the liberal majority in the antiwar movement.

Early efforts to do that took the form of ostracism, defaming protesting veterans as traitors and even communists. When VVAW members marched from Morristown, New Jersey, to Valley Forge, Pennsylvania, to protest the war in September 1970, older veterans of previous wars and stalwarts of Nixon's Republican Party base belittled them for their long hair and shouted for them to "go back to Hanoi!"[5]

In April 1971, VVAW staged an encampment on the National Mall in Washington to accompany its lobbying effort to end the war. John Kerry, representing VVAW, gave a passionate antiwar speech before a congressional committee for which he was later accused of betraying the security of the troops still in Vietnam. A year later, efforts to criminalize VVAW peaked when charges were brought against the

Gainesville, Florida, chapter for planning an armed attack on the 1972 Republican Party national convention in Miami Beach.[6]

DISCREDIT THEIR AUTHENTICITY

The widespread distrust of information flowing from Washington about the war meant that the voices from the ground level view of returned veterans were welcomed by the American public, but they were threatening to political and military elites. If those voices could not be suppressed or isolated, they would have to be discredited, their identity disputed, their authenticity impugned.

Beyond the crude suggestion that VVAW members might be agents of a hostile government and criminalized as seditious, critics suggested that protesting veterans were not "authentic," their numbers inflated by radicals posing as veterans. Speaking before a military audience in May of 1971, Vice President Spiro Agnew said he did not know how to describe the VVAW members encamped on the Mall but "heard one of them say to the other: 'If you're captured . . . give only your name, age, and the phone number of your hairdresser.'"[7]

In the same speech, Agnew said the antiwar vets "didn't resemble" the veterans "you and I have known," a statement, when combined with having gay-baited them, was an effort to draw an "us and them" distinction. Thusly drawn, the line invited more pronounced distinctions between "real men" and those who now refused to fight.

STIGMATIZING AS PATHOLOGICAL

Ostracizing and Otherizing are forms of stigmatizing, the denigration of an individual or group's identity. In his 1964 book *Stigma*, sociologist Erving Goffman wrote that the attribution of stigma can disqualify those "Others" from full social acceptance. In modern society, assignment of mental illness to targeted parties has become a powerful and pervasive form of stigmatizing.[8]

The first unlawful break-ins leading to the Watergate scandal of 1973 were those done by President Richard Nixon's "plumbers" on the psychiatrist's office of Daniel Ellsberg. A Marine Corps veteran and employee of the Rand Corporation in contract work for the government, Ellsberg had copied secret documents showing that political and military leaders had been misleading the public on the conduct of the war for years; the documents were later known as "The Pentagon Papers." Ellsberg had released the Papers to the press in 1971, incurring the wrath of the president. Wanting to discredit Ellsberg prior to the 1972 elections, Nixon assembled a team of former FBI and CIA agents to burglarize the doctor's office for files that would "destroy [Ellsberg's] public image."[9]

At the same time as the plumbers' raid on the psychiatrist's office, the press was hanging the same mental health markers on VVAW actions. When VVAW gathered in Miami Beach to protest the Republican Party's nomination of Richard Nixon as its presidential candidate in 1972, the *New York Times* featured a front-page story on the mental problems of Vietnam veterans. Beneath a headline reading "Postwar Shock Besets Ex-GIs" was a report peppered with words and phrases like "psychiatric casualty," "mental health disaster," "emotional illness," and "mental breakdown." The story acknowledged that there was little hard research on which to base those characterizations. Indeed, if the reporter had done his homework, he probably would have found Peter Bourne's 1970 book, *Men, Stress, and Vietnam*, in which Bourne, an Army psychiatrist in Vietnam, reported American personnel having suffered the lowest psychiatric casualty rate in modern warfare.[10]

After the 1972 *Times* story, the press tapped out a steady beat of stories about soldiers home from Vietnam with psychological derangements. In this, journalism was in step with the direction in which popular culture was headed with its representations of the war and the people who fought it. Hollywood had begun portraying Vietnam veterans as damaged goods since the mid-1960s and, consequentially, writing political veterans out of their stories. Films like *Blood of Ghastly Horror* (1965) and *Motorpsycho* (1965) anticipated the symptomatology of Post-Traumatic Stress Disorder before health care professionals coined the phrase. Controversies over the validity of war trauma nomenclature riled professional organizations throughout the late 1970s, and when PTSD was finally affirmed as a diagnostic category in the *Diagnostic and Statistical Manual of the American Psychiatric Association* in 1980, one of the authors of the terminology, Chaim Shatan, credited a *New York Times* opinion piece with having been the difference-maker in professionals' deliberations of its merit.

The shift from the political discourse that had dominated the veterans' homecoming story in the late 1960s and early 1970s to the mental health story line that became dominant in the 1980s is evinced in Hollywood film. In the 1978 film *Coming Home*, Luke (Jon Voigt) is politicized by his experience in Vietnam and shoddy treatment at a military hospital; he comes out publicly against the war. Four years later, we're given Rambo (Sylvester Stallone), who suffers flashbacks to the bamboo cage that confined him as a POW and then goes on a murderous rampage. Although political veterans like Luke were never prominent in feature films, the die was cast with Rambo. From then on, American filmgoers got a regular diet of warriors returning home with hurts, the unrepentant POW peacemakers among them.

More than the historical record itself, it would be the voices of veterans testifying to what they had done and seen that would shape public memory of the war and the POWs would be pulled into that agenda.

THE FORGOTTEN POW DISSENTERS

The POW story is as complicated and conflicted as that of the rest of the Vietnam-generation of veterans. The peace agreement that ended the war on January 27, 1973, stipulated the arrangements for the release of U.S. POWs, most of them held in prisons in and around Hanoi. The releases began on February 12 and continued in three increments until March 29. Carried aboard Air Force C-141 aircraft, they landed first at Clark Air Force Base in the Philippines for debriefing and medical assessment. From Clark, they were flown to stateside bases and on to regional hospitals and their hometowns.

News coverage of the POWs' arrival at Clark anticipated stories of dissent behind the bars that were yet to come. "Freed P.O.W. Asserts He Upheld U.S. Policy" read a February 15 *New York Times* headline before reporting that the pilot had made statements opposing the war while being held. A February 23 headline, "P.O.W.s Maintained Discipline but Had Some Quarrels and Were Split on the War," promised still more.

Intriguing as these headlines seem now, they may have been less so for readers in 1973. Peace activists and journalists had been journeying to Hanoi for years, where they met with POWs and heard a range of views about the war. George Smith, captured in the South and released in 1965, had written a book, *P.O.W.*, about his two years in captivity that was critical of the war. Far more interesting in retrospect is how widespread knowledge about POW dissenters in 1973 has been forgotten. As with the lost history of GI and veteran dissent recounted above, the story of POW dissent is less about forgetting than the reconstruction of what is remembered.[11]

"MUZZLED POWS . . ."

Not a salvo from an ACLU broadside, the headline "Muzzled POWs," replete with an ellipsis, topped a *New York Times* editorial—not an op-ed—on February 24, 1973. It followed a set of stories carried on its own pages about attempts to suppress news coverage of POWs' dissent since their release twelve days earlier. "P.O.W. Conduct Barred as Topic" read a February 5 headline a week before the first releases; "Managing the P.O.W.s: Military Public Relations Men Filter Prisoner Story" on February 20 told of eighty public relations specialists assembled to "hide possible warts and stand as a filtering screen between the press and the story."

Dissent had been a fact of life inside the walls of Hoa Lo for at least two years before the POWs' release and, not unlike their counterparts in leadership across the U.S. military system, the Senior Ranking Officers in the Hanoi prison system came down hard on those who deviated from the party line; efforts to suppress the protests included threats of courts-martial and even violence against them when they were released and returned to the States.

When outright suppression didn't work, the SROs warned others to stay away from the radicals, trying to isolate the bad apples. The prison administrators and guards controlled the movement and communications among the prisoners, of course, but the SROs contrived their own kangaroo command hierarchy into which they tried, first, to co-opt the highest-ranking dissidents, Navy Captain Walter Eugene Wilber and Marine Lieutenant Colonel Edison Miller, and, failing that, relieve them of military authority—"excommunicated" them, as historian Craig Howes put it. From then on, wrote Howes, "most men avoided them like the devil."[12]

The demonizing of the dissidents through "excommunication" is a recognizable form of ostracism, the same tactic as that implemented against the GI and veterans' antiwar groups in the States. By exiling the leaders, the SROs had set them up for victim-blaming. Their social exclusion would be construed by their peers (and later by the American public) as self-exacted, a kind of asked-for segregation, the responsibility for which lay with them. Putatively, the irrationality of their behavior stemmed from personal traits. They were loners, losers, alienated, and maladjusted, a cluster of shortcomings bespeaking weak character.[13]

The "weakness" notion was a form of slander, but it gained currency through its assignment to the ground troops captured in the South who arrived at the Hanoi complex in spring of 1971. Some of them were part of what was known as the Kushner group; many were younger enlisted men and lesser educated than the high-ranking pilots who preceded them into the Hanoi lockups; and the group was disproportionately non-white.[14] The ascription of personal failings to the Kushners' and other enlisted prisoners' antiwar leanings and resistance to the SROs' chain of command was a way to discredit the political authenticity of their opposition. Notwithstanding the condescension of rooting those "weaknesses" in the rebels' social backgrounds, the weakness language was also the rhetoric of character assassination. It dog whistled *moral* weakness, a failure to profess patriotic faith by willingness to suffer worldly deprivation. With a twist, it riffed on the mental health discourse already forming the narrative of GI and veteran dissent in which the voices of conscience, home from Hanoi, would fit.

The disrespect shown for the antiwar POWs followed them in their return home. From their first landing at Clark Air Force Base in the Philippines, through the White House welcoming staged for the POWs, their representation in the press, and further on in the stack of books that would be written about the "POW experience," they were the deviants whose behavior needed to be accounted for and explained.

FROM MUZZLED, TO CRIMINAL, TO MEDICAL: THE TRANSFORMING NARRATIVE OF POW DISSENT

Forgetting is an insidious process because it creates its own obscurity even as it takes place. In the case of the POW dissidents, the news about their censoring was short-lived, replaced first by stories about legal charges brought against them, and their defense against those charges. It was a kind of reversing-the-verdict maneuver whereby the stories about muzzled POWs that had the government on the defensive for the violation of freedom of speech were now reversed, putting the dissidents on the defensive for their conduct as prisoners.

Most news stories in March carried headlines like Seymour Hersh's for the *New York Times* on March 16, 1973: "Eight May Face Courts-Martial for Antiwar Roles as P.O.W.s." It was not as though the POWs were silenced by that spin so much as being compelled to speak as defendants in interviews with news reporters. This conflict was on display in Mike Wallace's interview with Gene Wilber on April 1. Confronted with Wallace's inference that he must have caved to the fear of torture—the "weakness" narrative that was building in the press—Wilber stuck with his claim to "conscience and morality." Wilber's stand effectively turned the tables again, putting the prosecutorial parties on the defensive for suppressing conscience.[15] Wilber's adherence to principle, however, was a grain of sand in the celebratory tide raising the stature of the "good" POWs reputed to have pridefully endured the "torture" handed out by their communist captors.

The POW news in May 1973 was dominated by President Richard Nixon's White House reception for them and their families. Press coverage of the reception totally erased the antiwar POWs from the story, and worse, did not cover the intimidations and threats exacted on them and their families behind the scenes.[16]

Legal maneuvering returned in the wake of the White House reception when Stockdale charged Wilber and Miller in late June. Wilber, through his appointed defense attorney, Lieutenant Commander Ronald M. Furdock, argued that the statements he made were not a violation of the UCMJ,[17] and that he and Stockdale had never met. Nor had Wilber met the "witnesses" listed as having knowledge of his "criminal" acts. Despite repeated attempts, Wilber's defense team was never permitted to interview the witnesses, some of whom also had never met Wilber. The allegations were rumors, hearsay that had been spread through the tap code.

The tap code was a Morse Code–like means of communication used by the POWs during periods of isolation. By tapping on the prison walls separating them, they maintained a bit of contact. Words were spelled by tapping two-character representations of a five-by-five matrix for each letter of the alphabet. Thus, row 1, column 1 was letter *A* (tap...tap); row 1, column 2 was *B* (tap…tap, tap). Messaging was slow, terse, and filled with abbreviations—not a forum for detail and nuance. Furthermore, messages were passed from room to room through adjacent walls from memory, a method that invited revisions and omissions, a form of "whisper down the lane" with no checks on accuracy. John Dramesi remembered the tap code as being a form of entertainment with lots of pornographic humor and gossip.[18]

Wilber and his attorneys were confident that the whispered communications in encounters in washrooms and limited communications through the tap code were the primary "witness" to his alleged crimes. These types of gossip communications were not going to stand up to the rules of evidence required in courts-martial, but they survive as the core of the "liberal dupe" slander that persists throughout the hero-prisoner narrative.

The likelihood of successful prosecution was slim. The Marine Corps had determined that "the case could not be successfully prosecuted."[19] The Commandant of the Marine Corps (regarding Miller) and the Judge Advocate General of the Navy (regarding Wilber) both recommended to Secretary of the Navy Warner that the charges against Miller and Wilber be dropped. Warner had made a three-month investigation of the charges before dismissing them; a court-martial would be disruptive to those who would have to testify, he said. In September 1973, he issued letters of censure to Wilber and Miller, charging them with failing to meet the standards expected of officers of the armed forces and announced they would be retired in "the best interests of the naval service." Ignoring that the government had a weak case that would not likely lead to conviction, Rochester and Kiley spun the dismissal of charges as an act of leniency extended to the "guilty" Miller and Wilber.[20]

By the time the legal case against Wilber and Miller collapsed, the dominant narrative had turned again. The rebels weren't criminal so much as emotionally and psychologically hurt, sick, damaged goods, just as were others of their ilk who had been in the streets as protesters against the war since 1967. A June 2 *New York Times* story headlined "Ex-P.O.W.s to Get Health Counseling for 5-Year Period," sprinkled in references to "high violent death rates," "depression," "fright," and "euphoria," with no references to sources for the claims. "Some Wounds Are Inside: Health of P.O.W.s" headlined a June 10 health column, raising mental health as the specter that would stigmatize dissent as a symptom. A July 15 story, "Antiwar P.O.W.s: A Different Mold Seared by the Combat Experience," locked in the mental health discourse despite there being virtually nothing in the content of the article to support the use of "seared" in the headline.[21]

The psychologizing of the dissenting views within the POW population was a way to dismiss their authenticity as political and moral expressions of conscience.

As it turned out, some never-say-die rightists who believed the war was not yet over, much less lost, were not about to have their disgust with POW dissenters salved with medical compassions. For them, Wilber, Miller, and the peacenik Eight were traitors—and there may have been more to what went on in Hoa Lo, and beyond, than what was known. Had everyone forgotten *The Manchurian Candidate*?

POWs OR MIAs?: THE NATIONAL LEAGUE OF FAMILIES BLENDS THE IDENTITIES

Throughout the late 1950s and early 1960s, there were a smattering of U.S. military and CIA personnel captured in Laos and Vietnam, but the purpose of their presence there was never clear, and they were all quickly released. The first captive who was on an acknowledged combat mission, as an "advisor," was taken in 1961. From then on, wrote historian Bruce Franklin, the number rose, but it was not until 1969, he says, that POWs became an issue.

According to Franklin, the revelations in 1968 that Americans had slaughtered civilians at My Lai and that the Phoenix program was arresting and assassinating suspected leaders of the National Liberation Front triggered President Richard Nixon to make an issue out of the way the communists were treating "our" prisoners. More pointedly, journalist Don Luce exposed in a February 1969 *Christian Century* article that tiger cages on Con Son Island held purported communists who were captured by the United States.[22] These exposés of American complicity in atrocities threatened to erode already sagging public support for the war as well as open government and military officials to charges of war crimes. As a counter-propaganda move on May 19, 1969, Nixon initiated what became known as the "Go Public" campaign to use U.S. prisoners held by the communists to leverage public opinion against the Hanoi government.

Nixon was also under pressure from the wives of POW pilots who had been frustrated by the outgoing Johnson administration's reticence on the POW issue. James Stockdale's wife, Sybil, was set in motion by a September 1, 1968, article in the *San Diego Union* newspaper titled "Red Brain Wash Teams Work on US Pilots." Mrs.

Stockdale had "pored over" books on the Korean War POW experience such as "*In Every War But One*," the Eugene Kincaid 1959 book that was an inspiration for *The Manchurian Candidate*.[23] Her reading had her worrying that government insiders were "too friendly with those Americans who happily spread North Vietnamese propaganda in the United States." She sent the *San Diego Union* article to Secretary of Defense Clark Clifford and Ambassador W. Averell Harriman and asked "what steps are being taken to prevent" these mistreatments of prisoners. When Harriman replied to her with assurance that the welfare of the prisoners was "uppermost in his mind," she later recalled her skepticism of his sincerity.[24]

Nixon's election in November opened a path for Mrs. Stockdale's access to the White House, and she took it. In July 1969, POW wives formed the National League of Families of American Prisoners in Southeast Asia with her as its head. The "League," as it became known, received funding support from the Republican National Committee, and RNC "advisors" helped coordinate its activities, according to Franklin.[25]

The League of Families lobbied the Nixon administration to negotiate the release of prisoners. The White House, however, embedded the prisoner release issue in its negotiations to withdraw troops from Vietnam and end the war. The U.S. military assaults would continue until the prisoners were released. The North Vietnamese were equally adamant—no peace, no releases.

The three-way stalemate over the POW issue continued until the last months of 1972 and was tightened by the League's demands that U.S. personnel missing in action (MIAs) be included in the negotiations. Questions arose immediately over who was known to be a captive and who was known to be missing—and *maybe* held captive. The wives of some pilots had been told their husbands had been shot down but were "missing." The absence of further information raised suspicions that the North Vietnamese were holding some pilots as prisoners but not admitting it. In fact, says Franklin, some of the missing were known by the Defense Department to have died in the shootdowns but their remains were unrecovered. The Defense

Department, he found, kept two sets of books: an undisclosed account that separated POWs from those known to be missing but BNR, or Bodies not Recovered, and a public account that lumped POWs and MIAs together.[26]

Unaccounted for or missing fighters are the sources of great anxiety for loved ones who have little more than their imaginations for filling in the blanks of the unknowns created by the separations. The unknowns also fuel the myths and legends that surround MIA issues. One of the greatest of those is the Legend of the Lost Command, born in the First World War's Battle of the Argonne Forest: a patrol went out and did not return. What happened to them? Were they lost? Were they captured? Were they dead? Had they deserted or defected to the enemy? Later it was revealed that a group in the unit had gone AWOL and ended up in German hands. When only one member of that group, an army private, returned, public suspicions raced to thoughts of treason and betrayal.[27]

The MIA soldiers, Marines, and pilots held in Vietnam were figures of no less fascination and, with Cold War intrigues factored in, speculation about the whats and wherefores surrounding them grew with time. Were the Vietnamese lying about who they held captive? Was Washington denying knowledge about some of the missing for reasons of mission security? How many, known to be dead, were still counted as POW/MIA by the Defense Department, a cruel use of information that kept their families in a painful state of suspended animation? Might some of the ground troops missing in the South have deserted—or even gone over to the other side, like the Peace Committee signees of Al Riate's proposal to do just that?

H. ROSS PEROT

The public's interest in the MIAs made it an immanently exploitable topic for political and ideological purposes. Just days after Richard Nixon's election in November, Texas multimillionaire H. Ross Perot funded full-page newspaper advertisements picturing children pleading for the return of their daddys being held in Hanoi. Perot was a

Naval Academy graduate, Second World War veteran, and entrepreneur in the then fledgling field of information technology. In June 1970 he funded a display on Capitol grounds to "arouse public opinion in behalf of the release of American Prisoners of War" and encouraged tourists seeing it to demand their release. In *M.I.A. or Mythmaking in America*, Franklin described the display:

> At the center were the figures of two American prisoners. One sits in the corner of a bare cell, staring bleakly at an empty bowl and chopsticks on which a huge cockroach is perched. On the floor are other cockroaches and a large rat. The other figure lies in a bamboo cage, ankles shackled. By the end of the year, this tableau was being set up in state capitols throughout the country.[28]

Franklin's description of Perot's exhibit as "simulated imagery" was apt, because few if any POWs were held in the conditions it portrayed. Its effect on public perception was to demonize the Vietnamese and their communist leaders and to displace the war itself, the actual war, from American memory. Missing as well was any hint that some American prisoners had by then, 1970, turned against the war and developed empathy for the Vietnamese. By that time, the mythology of flyers shackled in bamboo cages was already being monetized by the Victory in Vietnam Association.

BRACELETS BY VIVA

The Victory in Vietnam Association, known as VIVA, was chartered in California in 1967 as a tax-exempt educational organization to counter the effectiveness of the antiwar movement. It changed its name to Voices in Vital America in 1969 and began to sell metal bracelets with the names of POWs and MIAs engraved. Soon, says Franklin, VIVA was wholesaling bracelets through the League, Ross Perot's People United We Stand organization, and Junior Chambers of Commerce across the country. The bracelets were a moneymaking bonanza:

> Retailers kept 50 cents for each $2.50 nickel-plated bracelet and a dollar for each $3.00 copper bracelet, with the remaining two dollars going to VIVA whose costs averaged less than 50 cents per bracelet. . . . By early 1972, VIVA was distributing more than five thousand bracelets a day. . . . VIVA's income soared to $3,698,575 in 1972 and, despite the January 1973 peace accords, to $7,388,088 for 1973.[29]

Among the four to ten million Americans wearing the bracelets when the war ended were political figures Richard Nixon and George Wallace, actors Charlton Heston and Bill Cosby, the Dodgers' star pitcher Don Drysdale, and singers Johnny Cash and Sonny Bono.[30]

THE POW-MIA FLAG

The bracelets vaulted the POW-MIA issue to stratospheric emotional levels and lent political tonnage to the public's support of Nixon's pledge to keep the war going until "the last man comes home." But it was the POW-MIA flag designed and distributed by the National League of Families that became the most enduring symbol through which Americans came to remember the war in Vietnam. Flown over the White House on federal holidays every year since 1982, the flag shows a bowed black head with barbed wire strung beneath its chin with a guard tower looming in the background. The flag now flies daily above many state buildings across the country.

The Reagan administration throughout the 1980s maintained a *POW-MIA Fact Book* claiming "it would be impossible to rule out the possibility that live Americans are being held." The 1990 edition printed by the Bush administration said its POW-MIA efforts were "predicated on the assumption that some are still alive."[31]

The words "You are not forgotten," running across the bottom of the flag, disassociates those who fly the flag from those who would forget those left behind. It was a distinction that conjured weak-kneed Washington insiders who accepted Hanoi's denial that it was holding more prisoners than it admitted. The ending of the war on terms favorable to the North Vietnamese, said right-wing critics, confirmed that communist influence had penetrated the highest levels of national security—an echo of the Cold War conspiracism that had helped initiate the U.S. military mission to Southeast Asia in the first place. The soft-on-communism canard would remain a staple in the betrayal narrative for the loss of the war that would resound through America's end-of-century political culture.

ENEMIES: INSIDE THE BELTWAY AND THE WALLS OF HOA LO

American paranoia about communism goes back at least to the 1917 Bolshevik Revolution in Russia and the formation of the U.S. Communist Party in 1919. The CPUSA was instrumental in the rise of the CIO unions in the 1930s and a voice in early warnings of European fascism that kept it in the conservative crosshairs into the 1940s. Communism gained international stature through its role in the Second World War defeat of Nazi Germany, and communist-affiliated labor unions and political parties in Western Europe rode that popularity—and their own anti-fascist credentials—to new heights in the postwar years.[32]

Business and political leaders in the United States were unsettled enough by the Red tide arising abroad, but the specter of home-front communist influence was still more alarming. Fears that communist ideas were subverting religious and family values had the country on edge; speculation that communists had already infiltrated higher education and the entertainment industry set in motion efforts to purge the radicals. When suspicions arose of communists in the government, the country was overtaken by a mass hysteria that came to be known as McCarthyism.[33]

The communists were targeting the national security state, said Wisconsin senator Joseph McCarthy, and the bull's-eye was the Army itself. McCarthy's allegations led to the 1954 Army-McCarthy hearings during which the senator claimed to have a list of subversives in the State Department and the military. The hearings came at a time when emotions from the stalemated war in Korea were still settling and the country was grappling with rumors that some of its soldiers had crossed over to the communist North or Red China.

McCarthy's "list" never materialized, but the sensitivities raised by it carried over into the years of America's war in Vietnam. When GIs and veterans began coming out against the war and aligning themselves with left-wing organizations—said by rightists to be communist-front groups—Americans of various political stripes sensed a second coming of the McCarthy years. Post-1973 news that some POWs home from Hanoi had spoken out against the war while in lockup, combined with rumors that not all of the captured Americans had come home or were even accounted for, prolonged public curiosity about "what really went on over there." Dissenters among the returning captives, such as Wilber, Miller, and the Peace Committee, were now grist for imagining that they knew something about the missing and had even colluded with their communist jailers in the disappearance of their own comrades or, to give the conspiracy a twist, that they were colluding now with government insiders in a cover-up of the truth about the missing.

Privately funded POW-MIA rescue missions in the late 1970s came up empty-handed, as at Son Tay. Postmortems on the missions often pursued conspiratorial sell-out theories that led back to Washington—someone inside the Beltway had tipped off the communists that "we," the rescuers, were coming. By the early 1980s, films like *Uncommon Valor* were turning POW-MIA betrayal tales into box office gold.

JOHN McCAIN: ANOTHER MANCHURIAN CANDIDATE?

Although inspired by Hollywood, the most hardcore of the MIA betrayal tales emerged on the campaign trail. For a March 13, 2000, article in *Newsweek*, H. Ross Perot told the reporter that "he believes the senator [McCain] hushed up evidence that live POWs were left behind in Vietnam and even transferred to the Soviet Union for human experimentation, a charge Perot says he heard from a senior Vietnamese official in the 1980s." Perot added, "There's evidence,

evidence, evidence. . . . McCain was adamant about shutting down anything to do with recovering POWs."[34]

The story of POWs being traded to the Soviets had roots in Nelson Demille's 1988 novel *Charm School*, in which U.S. POWs have been traded to the Soviets. The Soviets are training Americans to be infiltrated back to the United States as agents. But were the traded Americans POWs or defectors? If defectors, at what point had they "gone over"? Years later, in Demille's story, the United States launches a raid. To rescue POWs? Or to assassinate defectors? The touch of nonfiction in the story's provenance is that Demille claimed in the book's preface that he had learned of pilots traded to the Soviets at Phu Bai airbase while stationed there during the war.

Two years later, freelance journalist Ted Sampley self-published his article, "John McCain: The Manchurian Candidate," claiming that Vietnamese communists tried to "turn American POWs into agents." Sampley cited the claim made by former Soviet KGB major Oleg Kalugin that "one of the POWs worked on by the KGB was a `high-ranking naval officer' who . . . agreed to work with the Soviets upon his repatriation to the United States and has frequently appeared on U.S. television." Sampley wrote that "Sen. John McCain fits [Kalugin's] description" and that the fact that he was a senator should not rule out the Manchurian Candidate possibility.

In June 1998, a stunning version of *Charm School* was broadcast on CNN as "Valley of Death," the product of months of investigative legwork. The report was that U.S. Special Forces had used Sarin nerve gas in "Operation Tailwind," a 1970 covert raid to assassinate a group of GIs who had defected to the North Vietnamese in Laos. Defected? Or were the targets really POWs that the U.S. government never wanted to come home? Had some of the war's MIAs actually been POWs that the government assassinated? The primetime broadcast, with legendary war reporter Peter Arnett narrating, only opened the question. A month later, CNN retracted the story and fired its producer, April Oliver.[35]

THE WAR IN VIETNAM FOR AMERICANS was very much about the communist leadership of the Hanoi government in the North and the National Liberation Front in the South. Although scholars and some politicians knew communism to be a social and economic system with features attractive to American workers—should they know the truth about it—government and business leaders used their power in religious and secular institutions to portray communism as a belief system based on lies, its followers dupes, victims of deceit and deception. Communism was godless, and Christian believers were called to the war against it.

As a war against mendacity with a religious tonality, the U.S. military campaign in Vietnam appealed to America's foundational mission as the "city on a hill," a beacon of divine light for a world vulnerable to the evils of the material world. The war had been a test of faith and the lures of draft resistance—"Girls say yes to boys who say no" went one antiwar slogan—dissent-in-uniform, and desertion, revealed the weaknesses of many young Americans. But the GIs, Marines, sailors, and flyers who resisted those temptations and fought the "good fight" came home with honor and went on to canonized heroism in "the official story."

6

A Captive Nation: POWs as Grist for the American Myth

A people's narrative is the story they tell about themselves, the story of how and why they came to be what they believe they are. The American story braids together religious and secular themes for a story about the "good people" threatened by powers that could destroy the group from the outside and the weakness of those on the inside who might betray the group. The religious strands of that story are as old as Genesis where Eve is unable to resist the tree of knowledge forbidden to her by God. From the Garden of Eden onward, the plight of humankind is a series of God's tests that winnow the weak from the strong, the true believers from the pretenders to grace.

Many of God's tests involve deception. In the New Testament, God's people are fooled into following the false prophets of Antichrist, the Beast, because he masquerades as good and promises relief from earthly miseries. Antichrist is revealed in governments and social policies that promise peace, justice, security, and happiness, when, in fact, these goals are achievable only through loyalty to God, not government. According to the prophets, God obligates his people to discern good from bad among those seeking earthly power. As a project begun by Puritans in the seventeenth century, young America was lulled with images of its place in the biblical narrative.[1]

JOHN SMITH AND POCAHONTAS

The first literature produced by colonial America was the so-called captivity narrative, stories written by Americans about themselves or others who had been captured by Indians. In a 1999 study of that genre, anthropologist Pauline Turner Strong wrote that American identity took form through representations of "struggles in and against the wild: struggles of a collective Self surrounded by a threatening but enticing wilderness, a Self that seeks to domesticate this wilderness as well as the savagery within itself." John Smith's story is one of the classics of the tradition.

Smith was captured on a reconnaissance mission into Indian territory, December 1607, six months after the founding of Jamestown. His captors were warriors of the Pamunkey tribe, led by chief Powhatan. Smith was held for several weeks, during which time he was spared from death by Powhatan's lovestruck daughter, Pocahontas. Smith negotiated his release, receiving land for Jamestown in return for his promise of cannons and a grindstone for the Indians. By this account, Smith saved Jamestown. Smith was a hero.[2]

Smith's story became the basis for his lionization in American history, but subsequent interpretations of what really happened during his captivity render it the stuff of myth. Strong points out that, even in the sixteenth century, the saving of Europeans by native princesses was already a literary cliché; she goes on to suggest that when Pocahontas "saved" Smith by casting her body between his head and the Indians' clubs, she was actually playing a role in a rebirthing ritual designed to transform Smith into a Pamunkey. In short, says Strong, Smith was probably submitted to an adoption ceremony, not an execution.[3]

The mythical qualities in Smith's version of his capture lie in the ambiguity it created about the firmness of the psychosocial boundary between him and the native Other who held him. In the anthropological light shed on his and stories like it, we're led to consider that the captivity experience may have confronted him with his own inner "savagery" and his vulnerability to Pamunkey paganism. Had Smith conjured the story of his near execution in order to disguise, even to himself, the appeal that conversion to "Indianism" had had on him? And was the hint of Pocahontas's seductiveness in the rescue scene a storytelling ploy to divert questions that might arise in Jamestown about his own attraction to the unbridled (as the English saw it) sexuality of the Indian girl and the impotence of the Puritan ethic as a restraint on him in that situation?

These questions were only implicit in Smith's story, but they gained prominence as new chapters were added to the captivity canon during the Indian wars at the end of the century. The narrative of temptation, punishment, and redemption that stories like Smith's framed became more complex after several instances of captives, apparently giving in to their inner Indian, rejecting the Puritan path and choosing to remain with their erstwhile captors. Those stories, in the hands of early eighteenth-century preachers, gave the captivity literature the qualities it needed for a meaningful interface with the POW narrative coming out of Vietnam some 350 years later.

KOREA, VIETNAM, AND THE RESUSCITATION OF AMERICA'S CAPTIVITY NARRATIVE

The captive Self and the captivating Other, as anthropologist Strong puts it, are the oppositional poles between which American identity is formed. That process of identity creation went on well into the nineteenth century as the ongoing conquest of North American Indians reproduced the need of Americans to see goodness in themselves and evil in others. The great and victorious wars of the twentieth century—the war to end all wars, 1917–1918, and the war against fascism, 1941–1945—did not tax that self-identity unduly, but the return of U.S. expansionism in the post–Second World War years did. A crisis of legitimacy that began with the war in Korea grew to dangerous levels during the Vietnam war years, requiring nourishment for the American images of Self and Other.

As if scripted by history, those wars in Vietnam and Korea also produced timely new generations of captives with stories that would reinvigorate the nation's collective narrative. Perhaps because the war in Vietnam followed so closely on the heels of Korea, it was the former that would make the greater contribution to the captivity literature. The facts of the war's widely perceived illegitimacy and its loss would, however, cause that narrative to turn inward with an intensity it had never had, spawning a search for the enemy within, and the imagining of that enemy in the absence of the real thing.

The new chapter of the "captive America" narrative that the POW experience of the Vietnam era would write, then, had, like the previous chapters, as much to say about America itself as its enemy.

GOOKS: THE ASIAN "OTHER" RACIALIZED

The representations of the POW experience in Southeast Asia did say a great deal about the enemy, of course, and much of it followed in the tradition of racial disparagement characteristic of the 300-year-old captivity stories. By the time the war began, moreover, that tradition had been given a specifically Asian twist by the war against the Japanese during the 1940s and the Korean War of the early 1950s. In both those conflicts, U.S. propaganda depicted Asians with slanted eyes, buck teeth, and sloping foreheads. In his book, *North Korea: Another Country*, Bruce Cumings recalls Koreans being described by *New York Times* military editor Hanson Baldwin as "barbarians," "primitive peoples," "armored horde," and "invading locusts." Other U.S. leaders and publications called them "half-crazed automatons" and "wildmen with arrested development." This "nauseating stew of racial stereotypes," wrote Cumings, stirred diverse peoples into a nameless sludge that accumulated under one name: "gook."[4]

The term "gook" carried over from the Korean War, becoming the appellation of choice for Americans referring to the Vietnamese. By the late 1960s, GIs arriving in Vietnam were quickly introduced to that word and its extended lexicon of "slants," "slopes," "zips," and "dinks" that diminished the people they were ostensibly there to support. Former POWs also adopted that language for their memoirs. In *Before Honor*, Eugene McDaniel describes one of his guards, measuring about five feet ten inches, as the world's tallest "gook"; about another he wrote, "savage, and that said it all." Larry Guarino wrote without apparent sarcasm that Vietnamese "brains functioned on different frequencies from ours . . . so I gave up any thought of reasoning with them and declared them to be 'just gooks.'" Ralph Gaither recalls being driven from Camp Dogpatch to Hanoi by "a bunch of gooks." [5]

Racist portrayals of the Vietnamese appear in most of the POW memoirs, and in some instances the writers use descriptions of deviant behavior and local customs to make their point. Gaither, for example, described what he saw from Dogpatch:

> The guards made horrible sport of butchering their food. When they killed a pig, they first punched its eyes out so that it would stand still, then they beat it to death. Cows suffered a like fate. They tied them with ropes strung all over like a spider web, and then beat them to death with an ax. The women killed the dogs for meat by hanging them in a grotesque position and then beat them with sticks. All the while the guards and women laughed as though they were at a sports event.

From that, he concluded, "You could not generate much respect for such people as that. We prisoners could not expect much better treatment than they gave their animals and that's about what we got."[6]

One may take at face value the accuracy of Gaither's report and still question his use of it to describe the treatment of POWs. Cruelty to animals might predict cruelty to people, but we would have to know more about the cultural practices of the people and circumstances surrounding the events he witnessed before accepting the analogy he constructs. Using the treatment of animals as an indicator for the national character entails an even greater leap. Even though U.S. treatment of prisoners was as bad or worse as that dished out by the

North Vietnamese, one could make little sense out of that record by having witnessed, say, U.S. soldiers shooting elephants from helicopters for sport. For that matter, Vietnamese prison guards are likely to have been a small and select portion of the population, from which, whatever their behavior, it would be dangerous to generalize about the nation's people.

THE CAPTIVE SELVES IN HOA LO PRISON

The enduring presence of the terrifying oriental Other in the Vietnam POW story is based less on America's continuing need to believe the worst about the Vietnamese than on the way that image helps construct what Americans want to think about themselves. Not only are "we" not like "them," our POWs proved that Americans would not forsake themselves or their nation even under the extreme conditions of imprisonment by the enemy. The standard account of the American POWs in this narrative is that the abrupt cessation of torture in late 1969 was due to the death of the devilish Ho Chi Minh, or to the early release of a few prisoners whose public testimony on torture upon return to the States sparked a letter-writing campaign by the POWs' families that put pressure on the Hanoi government to lighten up—an explanation supportive of the idea that the struggle of the prisoners, and their families by extension, was meaningful.[7] By this reasoning, the Vietnamese were a racial Other lacking the civilized sensibilities that peacenik naïveté imparted to them. Acceptance of this version, moreover, required belief that the prisoners were masters of their own fate, which made their improved conditions and eventual repatriation explainable as results of their own struggle, not of conciliation and peace.

But there is reason to wonder if the greatest change in 1969 was in Vietnamese prison practices or in the story told by the SROs about the pre-1969 conditions. Was it just a coincidence that conditions changed at Hoa Lo and other Northern camps at the same time the first large group of GIs captured in the South arrived there, or was their arrival the trigger for a change in the story the SROs needed to be told? Is it possible, in other words, that the disparity between the pre- and post-1969 conditions looks so great because the SROs constructed a kind of "bad ol' days" mythology that exaggerated the tough times they had been through in order to impress and establish authority over the newly arrived prisoners?

The suggestion of a political subtext to the SROs' story implies that there was a game of one-upmanship being played by the wannabe heroes who were sensitive to the weakness of their own credentials as tough-guy holdouts. The competing interpretations to be made of the rough interrogations that the captives were put through—was it torture in pursuit of militarily significant information or punishment for misconduct?—are still in play.

SELF-FLAGELLATED HEROES

The most fascinating renditions of the torture stories come from POWs who openly admit to having consciously provoked the Vietnamese to torture them, and at the time even inflicting physical injury on their own bodies, essentially torturing themselves when the Vietnamese wouldn't do it. Like Everett Alvarez, other pilots captured during the first year of the air war quickly recognized how little was at stake in the interrogation sessions and gave the Vietnamese useless information in order to avoid physical coercion. That changed, however, with the arrival in Hoa Lo of Navy Commander Jeremiah Denton.

Denton came in July of 1965 and declared himself in charge of the thirteen other captives, some of whom had been there for nearly a year. "Follow the Code," he ordered, and plan for escape and resistance. As it played out, Denton's order was a dangerous act of provocation by which, he admitted to the press years later, "We forced them to be brutal to us." There was a sadistic inflection in that approach, moreover, because Denton ordered all those POWs who had already "failed," by cooperating with the interrogators, to "bounce back" by confessing their weaknesses to him or another SRO and then return to interrogation sessions where they could then "succeed" in resistance and have another chance to endure torture. The threat for not repeating the cycle was that SROs might bring charges, years later after release, against the insubordinate underlings, who had been "broken" by the Vietnamese. Had that part of Denton's order been put into words for the press, it might have read, "We SROs forced the North Vietnamese to be brutal to our lower-ranking fellow POWs."[8]

James Stockdale and Robinson Risner followed Denton into the prison system a couple months later. For them, as for Denton, torture wasn't something to be avoided, but a test of will and faith to be passed. Reversing the conventional logic whereby torture was the consequence befalling the prisoner for failure to give the Vietnamese the information they wanted, the three leaders posited war as the normative relationship between the POWs and their captors, with torture being the form that combat took in those circumstances. For these hardliners, the cycle connecting interrogations and torture began with torture: to be "at torture" was the natural state-of-being for the American POW. Like any form of warfare, torture was painful, it took men to their limits, and required periodic retreats, that is, a return to the interrogations. But warriors bounce back, they return to combat, and the prisoner *at* war uses the interrogations as the road to take back to the front lines where he belongs—in the stocks, under the lash, and in the ropes.[9]

The fact that senior officers among the POWs actually provoked mistreatment by the Vietnamese has always been an asterisk on the torture stories that POWs told upon their return. But, as a segue to the religious subtext that makes the Vietnam-era POW stories so resonant with the early American captivity narrative, the accounts coming out of Hanoi of self-inflicted bodily pain are still more poignant. In January 1969, for example, Stockdale began fasting to protest Vietnamese attempts to extract what even he knew was useless five-year-old information from him about his unit. Then, fearing that they might try to film him for propaganda purposes, he battered his own face with a stool and cut his scalp to create wounds that would appear to have been inflicted by his captors. Stockdale's self-mutilation became a kind of theater that he staged in order to assuage his anxiety and self-doubt, with his most effective prop being the "puke balls" he made by swallowing soap and then vomiting. This abuse of himself went on for days, with self-delivered "early morning bashings" that turned his face and hands into mashed flesh. "It really became a test of self-discipline," he recalled later.[10]

Risner in his book told a similar story. Fearing he might be photographed or filmed for propaganda purposes, he said he first considered killing himself and then thought about cutting the tendons in his hands. He doesn't say what purpose crippled hands would have served in that context, leaving the reader to surmise that he added it to his story to conjure images of self-administered crucifixion. In any case, Risner eventually reasoned that it was his voice that the Vietnamese really wanted. Since he was a high-profile captive, an American hero even in the eyes of his fellow inmates, his voice heard reading news or information over Radio Hanoi would lend credibility to North Vietnamese broadcasts. So it was his voice that he would deny them.

Recalling that drinking acid or striking the throat could damage the voice, Risner knelt and prayed. "I said, 'Lord, is this the right way to go? Should I cut my wrists? Should I try to destroy my voice?'" With or without divine guidance (he doesn't say), Risner went for the throat:

> I began pounding my throat as hard as I could. My eyes watered and sometimes I saw stars. I gave frequent chops to the throat. . . . I choked and struggled to get my breath. My neck swelled up, but it did not affect my voice. I could still talk.

Risner turned to the acid treatment, counting on his soap to have enough lye in it to damage his throat:

> I took my cup and filled it with a third cup of water, and part of a bar of lye soap. I crumbled, mashed, and stirred it into a mushy, mucky substance. Then I began gargling with it. It burned the inside of my mouth like fire, almost cooking it, and I accidentally

> swallowed some. The taste was enough to make me vomit. To increase the effect of the acid, I decided also to try to damage my vocal cords. I held a rag or towel over my mouth and screamed as loud as I could but at the same time compressing the air I was expelling so there would not be much noise. I continued this for three days and nights, staying awake as much as possible. By the third day, I could not whisper. I tried to talk, then tried to sing. I could not do either. It had worked! I was now a mute.

It didn't work, and Risner wasn't a mute; one good cough the next day and his voice was back, leaving him with nothing but diarrhea for his effort.[11]

For Denton, Stockdale, Risner, and other hardliners, torture became a way to confirm their worth as American warriors. Evidence of physical damage from torture, they hoped, would be evidence, upon their release, that they had remained *at* war during their imprisonment. When the torture they wanted from the Vietnamese wasn't forthcoming, they provoked it. When that didn't work, they inflicted their own damage. As Stockdale put it in his memoir, he prayed that his binge of self-flagellation would "make Syb [his wife, Sybil] and those boys of ours proud."[12]

The story of prisoners *at* war is the "official story," but it is the story of only a few of the hardcore SROs and it isn't always clear that the war being fought within Hoa Lo was the same war being fought outside the walls. It appears upon reexamination, for example, that SRO demands that prisoners resist interrogations had little to do with military strategy and a lot to do with SRO control over the prison population and the twisted sense of self-discipline for which some SROs like Stockdale needed to have validated by being tortured. Likewise, the resistance of senior officers to what they considered their exploitation for propaganda purposes probably had more to do with the war in America than the war in Vietnam. Robinson Risner's futile effort to disable his voice lest he be forced to read something over Radio Hanoi that the other POWs might hear is a case in point.[13]

Risner doesn't say what it was that he would have had to read, nor does he date the time of his struggle with that issue, but it occurred about the time of a similar incident described by Stockdale in his memoir. Stockdale objected to prisoners being used to "read the news" because he had learned in a course at Stanford University, "Comparative Marxist Thought," taught by "an old Kremlinologist who knew all [the Marxist strands], that the use of prisoners that way could be manipulated for brainwashing." The point had been driven home for Stockdale by tapes made by psychologists who had studied POWs from the Korean War.[14]

But is that what was going on in Hoa Lo? By Stockdale's own account—dismissed by him as communist trickery—the Vietnamese wanted an American to read the news because their interpreters did not have sufficient skill in English to be understood by the Americans. And what was the news they wanted read? Stockdale recalled it being *New York Times* excerpts from the paper's assistant managing editor Harrison Salisbury, who had recently returned from his fact-finding trip to Hanoi. True, Salisbury had come back to the United States critical of the Johnson administration's Vietnam policy, but if there was a credible source to be had from the U.S. side of the war, he was it.[15] Stockdale nevertheless put out an order that participation in the readings should stop, and twenty years later, in his memoirs, he stuck to the Cold War silliness that Salisbury's *Times* report was "loaded with what we had all been getting for months as the [Communist] Central Committee's propaganda line."[16]

Taken at face value, Stockdale's denial of the news to the POWs appears due to his fear that the ideological war beyond the walls of Hoa Lo was more than his men could handle. So it was best to protect them from it. Led by the reinterpretations of other captivity stories made by Strong and Gruner, however, we have to ask now if what Stockdale feared more was the enemy inside the walls, indeed the enemy within the hearts and minds of the POWs themselves.

The SROs' fear that their underlings might defect, and evidence for the legitimacy of that fear, can be found in the memoirs. By the early 1970s, there was open rebellion against the SRO leadership. As discussed in chapter 3, some of the discontent was fomented by the newly arrived POWs from the South who had survived jungle confinement for months and years before arriving in the Hanoi system. Their experience made them skeptical of SRO "authority" that was based on the officers' claims to having suffered abuse between 1965 and 1969.[17]

THE POWs IN THE AMERICAN CAPTIVITY NARRATIVE

There was more to antiwar dissent than struggle over the distribution of power among the inmates, but that issue is a clue to understanding the Vietnam-era prisoner experience as an extension of the American captivity narrative. The resistance of rank-and-file POWs and dissident officer-pilots like Wilber and Miller to the control of the SROs, who were committed to their own place in a post-release hero-prisoner script, was accompanied by emotional sentiments that aligned them with the Vietnamese, the enemy Other as seen by hardliners.

It was the humanizing of the Other that the SROs feared because recognition of the humanity in "them" meant recognizing an element of them in "us," an acknowledgment of the enemy within the hearts and minds of GIs, of Marines, of fellow pilots.

With the abundant inconsistencies in the accounts of torture handed out by the historians John Hubbell, Stuart Rochester, and John Kiley, and the numerous references in the POW memoirs to SRO fears of the "white gook" within themselves or their comrades, it seems likely that, as much as for John Smith, it was the *attraction* of the Other that scared the SROs. It was enough to mask their vulnerability with a counternarrative about the terror that the Vietnamese visited upon them. In short, it's possible that it wasn't the extinguishing of themselves they feared—execution in the Smithian legend—so much as their conversion to the ways of the Vietnamese and the righteousness of their cause.

Reflecting on his own interrogation experience, Army captain John Dunn turned over the same stone that obscured the Pamunkey side of history for centuries when he wrote that his prison administrator "seemed more interested in conversion than . . . the application of torture to obtain information." Dunn's insight wasn't just imaginative. Many of the POWs described the education programs the Vietnamese provided for them, the content of which seems to have differed little from what was being taught in U.S. college classrooms by the late 1960s: the history of Vietnamese struggles for independence, the details of the Geneva Accords that had ended the French occupation and artificially divided the country, and basic political economy. From what we know, there was little in the Hoa Lo "lesson plans" for POWs that is not today the accepted wisdom of what the war was about. That being the case, the claim by one of the Vietnamese interrogators that the prisoners were being given "the right to rebirth," seems all the more plausible, all the more sincere, all the more resonant with take-two on the John Smith legend.[18]

7

The Heritage of Conscience: From the American War in Vietnam to America Today

The in-service resistance carried out during the war in Vietnam by airmen, Marines, GIs, and sailors was documented by David Cortright in his 1975 book *Soldiers in Revolt* and brought to the screen in 2006 in David Zeiger's film *Sir! No Sir!* Cortright and Zeiger's contributions to American history and public memory of the war are, however, exceptions in the volumes of scholarly work and popular culture with which most Americans are familiar. All too often, those better-known accounts leave out the story that some of the warriors sent to fight the war also fought gallantly against it. The fact that POWs held in Hanoi numbered among those dissenters has been all but erased from the record.

CARRYING IT ON

The heritage of service members and veterans who acted with conscience to help end the war in Vietnam was kept alive in the decades that followed by the men and women who had formed Vietnam Veterans Against the War (VVAW). Throughout the 1980s, VVAW rallied against U.S. interventions in Central America, often forming the lead contingents for protest marches in Washington, D.C., and New York City. And individual veterans, who may or may not have been members of VVAW, provided some of the most powerful profiles of service members' lifelong commitments to peace and postwar reconciliation.

CHARLIE CLEMENTS

Charlie Clements graduated second in his class at the Air Force Academy and flew missions in Vietnam before citing moral grounds in refusing to fly anymore. Remanded to a military hospital for psychiatric evaluation, he left the military, became a doctor, and inspired a new generation of activists working to end American wars abroad. His book and Oscar-winning 1986 film *Witness to War* documented the indigenous resistance to the U.S. puppet regime in El Salvador and made known to a broader public the work done by Medical Aid to El Salvador, an organization he helped found.

JAN BARRY

Sent to Vietnam in 1962 as a radio technician, Jan Barry felt an immediate connection with the Vietnamese who worked on his base in Nha Trang. Appointed to West Point upon returning stateside, he grew increasingly unhappy with the war, and dropped out after the U.S. launched air strikes against North Vietnam in 1964. Barry joined the Spring Mobilization to End the War in Vietnam in April 1967 and in June helped draft the bylaws for Vietnam Veterans Against the War. Today, Barry is an acclaimed poet, considered by writer and critic W. D. Ehrhart to be "the most important figure" in the Vietnam generation of poets.

CHUCK SEARCY

Searcy was an intelligence analyst based near Saigon in 1966. The information he gathered led him to question the truth of what the government was telling the American people about the war. Upon return, he graduated from the University of Georgia and edited a newspaper in Athens, Georgia. In 1992 he returned to Vietnam to help locate and defuse unexploded bombs dropped by the United States that continued to kill and maim Vietnamese farmers. After international acclaim for his work with Project Renew, an effort to locate and disarm unexploded ordnance in Quang Ngai Province, Searcy, now in his seventies, continues his mission of peace.[1]

SUSAN SCHNALL

After graduation from Stanford University's nursing program in 1967 and commissioning as a U.S. Navy officer, Susan Schnall was sent to Oak Knoll Naval Hospital in Oakland, California, where she took care of casualties from Vietnam. Made aware of a GI and veterans march for peace in the San Francisco Bay Area in October 1968, she got involved and organized her corpsmen and WAVEs to participate. They put up posters in the hospital at night that were torn down by the morning. Thinking she could do more, she contacted a friend who was a pilot, filled his single-engine plane with fliers about the upcoming demonstration, and dumped them from the air over five military installations in the San Francisco Bay Area. On October 12, Lieutenant Junior Grade Schnall participated in a demonstration in her uniform for which she was court-martialed.

When charges were dismissed, Schnall moved to New York to do clinical work and organize for health care reform, including, at the time, Medical Aid for Indochina, which raised money for medical supplies that went to the North and the National Liberation Front. Since 2006 she has led efforts to aid Vietnamese victims of Agent Orange, and she also works with Veterans for Peace (VFP). Reflecting on her life of activism for an August 22, 2019, interview, Schnall mused on the importance of people knowing they can take a step against authority and say, "I disagree with you, and I'm going to do something about it." And by doing that "you'll survive with your moral conscience intact."[2]

THE HERITAGE ORGANIZED: VIETNAM VETERANS AGAINST THE WAR TO IRAQ VETERANS AGAINST THE WAR

The spirit of peace born in Vietnam carried into the twenty-first century. Following the invasion of Iraq in the spring of 2003, a new generation of veterans attending the 2004 VFP annual convention formed Iraq Veterans Against the War (IVAW) with VVAW as an organizational model. In March 2008, IVAW conducted hearings that exposed atrocities committed or witnessed by its members in Iraq, a whistle-blowing exercise modeled after VVAW's Winter Soldier hearing in 1971.[3]

The organized expressions of dissent manifesting as IVAW is the most significant legacy of the Vietnam generation's dissent. But it is easy to imagine that others, inspired by their dissenting forebears and acting as individuals, stepped up to protest the new wars in the Middle East. Matthew Hoh, Ann Wright, and Pat Tillman were among them.

MATTHEW HOH

Matthew Hoh took part in the American occupation of Iraq in 2004–5 with a State Department reconstruction and governance team, and then in 2006–7 in Anbar Province as a Marine Corps company commander. On State Department assignment in 2009, Hoh resigned in protest of the Obama administration's escalation of the war against the Taliban. In 2010 he was awarded the Ridenhour Prize Recipient for Truth Telling named for Ron Ridenhour, the Vietnam veteran who exposed the 1968 My Lai Massacre. Today, Hoh is on the boards of Veterans for Peace and World Beyond War and is a frequent contributor to American news media. He has been a source for the *Washington Post* and *Wall Street Journal* and has been a guest on many network and cable television news programs.

ANN WRIGHT

Ann Wright retired as a colonel after thirteen years in the Army and sixteen years in the Army Reserves. She went to work for the U.S. State Department Foreign Service in 1987 and facilitated the opening of the U.S. Embassy in Afghanistan after the 2001 invasion of Afghanistan. Wright resigned in protest from the State Department the day before the 2003 invasion of Iraq. Wright has been arrested many times for protests of U.S. occupations in the Middle East and was one of three witnesses called to testify in June 2006 in support of U.S. Army Second Lieutenant Ehren Watada, who refused to deploy to Iraq with his unit, asserting that the war there violated the U.S. Constitution and international law.

PAT TILLMAN

Pat Tillman was a standout linebacker at Arizona State University, picked in the football draft in 1998 by the Arizona Cardinals. Profoundly moved by the attacks of September 11, 2001, he turned down a three-year contract worth $3.6 million and enlisted in the Army. In spring 2003, he deployed for the invasion of Iraq, disappointed to not be sent to Afghanistan in pursuit of the alleged 9/11 perpetrator, Osama bin Laden.[4]

The apparent conflation of U.S. leaders' missions in Iraq and Afghanistan was Tillman's first inkling that there was more to the American agenda in the Middle East than met the eye. Tillman's biographer Jon Krakauer, with access to Tillman's personal journal, noted his disillusionment with the war in Iraq. In words echoing Vietnam War POW dissidents James Daly or Bob Chenoweth, Tillman had written: "You know, some of these [Iraqi] kids are getting to me. . . . There are some very good people, especially some of these kids." With time, his feelings against the war would harden.

By May 2003, Tillman was complaining about leaders "telling guys to shoot innocent people only to be ignored by privates with cooler heads." By the time he redeployed to Afghanistan, he had become interested in Noam Chomsky's ideas and was looking forward to a meeting with the professor when he returned home.[5]

On April 22, 2004, Tillman was killed in a friendly fire incident. Evidence that the deadly rounds were fired within ten yards of him, and revelations that his uniform and his recent journal with his thoughts about the war had been burned at Forward Operating Base Salerno from which his unit worked, spawned rumors that he had been assassinated. Krakauer later dismissed the assassination idea as conspiratorial speculation.[6]

THE BLOWBACK TO CONSCIENCE

If the heritage of the conscientious rejection of war exercised by the Vietnam generation of GIs, POWs, and veterans reemerged in the biographies of Ann Wright, Matthew Hoh, Pat Tillman, and the veterans who organized IVAW, it would be an inspiring story certain to rouse the passions of war resisters going forward. But, as with the Newtonian principle that every action comes with an equal and opposite reaction, the steps taken toward peace by the dissidents locked in Hoa Lo and conscientious peers on the outside called forth the reaction of authorities and rejection by a Cold War–fearing public. Out of that push-and-pull, the betrayal narrative for the loss of the war forged discourses for political and ideological struggle over the U.S.'s twenty-first-century role in the world.

Units deployed to the Middle East were made up of volunteers, many of whom were hot to avenge the attacks of September 11, 2001. They arrived at Ayn al Asad Airbase in Iraq or Camp Phoenix in Kabul without the edges carried by draftees to Vietnam; military contracts with Burger King and other chains had homestyle comforts waiting for them.[7] Co-opted though they may have been by the relative comforts of new-century warfare, the ages of the troops—twenty-six was the mean age of the Americans killed in Iraq—and the educational backgrounds of many, like Pat Tillman, predisposed them to skepticism about the mission and its leaders once they saw it for real.

DISSENT PATHOLOGIZED AND MEDICALIZED: "VICTIM-VETERANS" RE-UPPED

Simultaneous with the nudge toward resistance given by Hoh, Wright, and the IVAW leaders by the examples set by VVAW, dissent was discouraged and discredited.[8] Indeed, the coming-home story of the Iraq War veterans had been scripted before they left home. The dispatch of troops to the Persian Gulf in the fall of 1990 had drawn opposition from the antiwar movement. Prowar conservatives tried to discredit the movement by claiming it involved the same people who had spat on Vietnam veterans and betrayed the American mission in Southeast Asia. The Yellow Ribbon Campaign followed, and when troops returned in the spring of 1991, the news media lit up with reports of traumatized returnees returning to a public disinterested in their welfare.

The victim-veteran imagery—a carryover from the war in Vietnam—was tagged twelve years later by writers looking for ways to narrativize the new war. Joseph B. Verrengia, an Associated Press science reporter, prototyped the model that would shape news coverage of Iraq War veterans for years to come: "How many soldiers will require mental health treatment?" he asked in an April 18, 2003, story. Traumatized soldiers, he continued, "relive their horrors through flashbacks and nightmares, often followed by depression and fury." Moreover, he wrote, "this war [in Iraq] is colored by controversy and protests."[9]

Virtually every major newspaper would produce a feature story on "wounded warriors," many of them focused on the mental and emotional damage of the war. One of the earliest and most powerful was the *Boston Globe*'s four-part series, "The War After the War," begun in October 2006.[10]

The victim-veteran discourse so dominated the news of homecomings that dissent was squeezed out of the conversation. Thomas

Barton told Jerry Lembcke that when IVAW organizers visited New London, Connecticut, in 2006, the public's preoccupation with PTSD had made it harder to organize. "Everywhere we go," he said, "all people want to talk about is PTSD." The identity of veterans empowered and politicized by their wartime experience seemed overwritten by that of men and women home with unseen hurts asking for sympathy more than solidarity.[11]

DISSENT ABOARD THE CARRIER USS *THEODORE ROOSEVELT*

As much as the conscientious dissent of the Vietnam-generation fighters and veterans played out in the new wars of the twenty-first century, it was met by an opposing reaction that also was descended, and inherited, from the same era. Indeed, the rise of the Trump movement in the early twenty-first century can best be understood as an extension of the betrayal narrative for the loss of the war in Vietnam. The Trump slogan, Make America Great Again (MAGA), posits a largely mythical prelapsarian America whose greatness was eroded by the permissiveness and economic entitlement programs associated with the 1960s.

But maybe that polarization is but a new beginning. The same push and pull, viewed by the German philosopher Hegel, is a *creative* energy leading not to a renewed stasis but a new, and higher, level of struggle. During the 2020 coronavirus pandemic, the virus broke out onboard the deployed aircraft carrier USS *Theodore Roosevelt*. The ship's captain, Brett Crozier, called for help from his chain of command in combatting the spread of the deadly virus among his crew. When his pleas went unheeded, he put the welfare of his sailors ahead of his own career and went around the usual channels with a letter to still higher ranks—an exercise of conscience befitting the dissident POW officers who sacrificed their careers to help end a war that was taking the lives of fellow fliers and shipmates.[12]

President Trump made known his displeasure with the captain, and on April 2, acting Navy secretary Thomas B. Modly fired Crozier.[13]

We don't know if the examples of Gene Wilber and Ed Miller were influences on Crozier. Cultural influence is elusive, intangible, touching the emotions as well as the mind. Maybe he encountered the history of the Hanoi dissenters in a class at the Academy, or maybe he just heard about it from classmates. Or maybe he saw *The Hanoi Hilton* movie. We don't know. Historians, in any case, make a distinction between objective and subjective social forces. It is a distinction that makes it less important that Crozier was moved by knowing that the officers held in Hanoi broke Navy protocol, resisted self-appointed authority figures, and stuck to their consciences. Maybe he was moved by something in his own background, social relations on board the carrier, and the organizational and political cultures of the times, as he saw them—objective social forces. We don't know.

The temptation to put Crozier's act of conscience on a trajectory tracing back to Hanoi grows stronger, however, when we see the same character displayed by his crew members. When Crozier departed *Theodore Roosevelt* after being fired, the crew gathered for a supportive mass farewell, chanting his name in respect, a rejection of the authorities who had cashiered the skipper for putting their well-being ahead of his own. A few days after Modly fired Crozier, he flew halfway around the world from inside the beltway, boarding the *Roosevelt* to justify his decision to the crew. Over the ship's intercom, Modly asserted that the captain had been canned because he was "stupid and naïve," to the apparent astonishment and ire of the crew. A recording of Modly's rant against Crozier taken from a loudspeaker in one of the many crew spaces also captures comments by listening sailors: "What the fuck?" "He was only trying to help us!" "Oh?" along with other muttered groans questioning Modly's characterization of this matter of courage and conscience.

Still, the conscientious actions in the Crozier affair and the crew's reaction could be an anomaly, an exception to military obeisance to authoritarian political figures. An example is the pre–Second World War German military that fell in lockstep with the Hitler cult for a fascist crusade to restore German greatness allegedly lost in the First World War. It is that pattern in the historical record that unsettles some observers of the United States today. Might the threads of white supremacy and nationalism in the MAGA movement have been vested by Trump with the martial might to wrench American lost-war angst into something as dangerous as Germany's interwar revanchism?[14]

Ironically, the kind of servile military leadership that Trump had in mind might itself have been a casualty of the war in Vietnam. The May 18, 2020, The *New York Times* reported findings of a *Military Times* survey that "50 percent of active-service military hold an unfavorable view of the president." The *Times* quoted reporter Mark Bowden saying, "I have never heard officers in high positions express such alarm about a president," and called the Trump presidency a "slow-motion train wreck in civil-military relations."[15]

Fifty years after the war in Vietnam, the acts of conscience displayed there, and the reaction they provoked, continue to drive American political culture. The struggle over the heritage of that experience, waged between those who tell, interpret, and decide the uses to which it is put, looms as large as ever in the meaning of the war in the nation's present.

Notes

Introduction

1. In his essay, "Missing in Action in the 21st Century," H. Bruce Franklin outlines how Perot's Christmas package trip fit into the overall "Go Public Campaign" begun by the Nixon administration in March 1969 to stall the peace negotiations with the POW issue, https://www.hbrucefranklin.com/articles/missing-in-action-in-the-21st-century/.
2. Mary Hershberger, *Traveling to Vietnam: American Peace Activists and the War* (Syracuse, NY: Syracuse University Press, 1998), 177–200, has a detailed account of the Committee of Liaison's mail service.

1. Forgotten Voices from Hoa Lo Prison

1. The Code of Conduct (Article III) refers to "parole" and "special favors" as examples of things that a prisoner should not accept from their captors. Vietnam POW literature refers to a prisoner's acceptance of early release, or amnesty, as a violation of this article. Better food or medical care than given to their peers (interpreted as special favors) would be a similar violation.
2. The validity of the torture allegations will be explored in the following chapters. Relatedly, there are questions about the motivations for the harsh treatment, whatever its measure, that was meted out by the prison staff: was it punishment for breaking prison rules, a "brainwashing" tactic, or an effort to extract militarily sensitive information? According to historian John G. Hubbell in *P.O.W.: A Definitive History of the American Prisoner-of-War Experience in Vietnam, 1964–1973* (New York: Reader's Digest Press, 1976), Navy pilot James Stockdale,

shot down in September 1965, was still being tortured in 1969 for information on aircraft air defenses, which is unlikely since any information obtained would have been outdated and useless to the Vietnamese (476).

3. Now used by establishment scholars to valorize its quality and legitimacy, and sardonically by critics, "official story," according to Stuart Rochester and Frederick Kiley in *Honor Bound: American Prisoners of War in Southeast Asia, 1961–1973* (Annapolis, MD: Naval Institute Press, 1999), xi, was coined by English Professor Craig Howes in his *Voices of the Vietnam POWs: Witnesses to Their Fight* (New York: Oxford University Press), 75.
4. Hubbell's *P.O.W.* acknowledgments give thanks to *Reader's Digest* editors Andrew Jones and Kenneth Y. Tomlinson for their help with the research. Tomlinson's Wikipedia entry details his right-wing ties with the Reagan and Bush presidencies and charges made against him for the "propagandistic" quality of his work in the media. The Wikipedia entry credits him with co-authorship of *P.O.W.*, while another site credits Hubbell, Jones, and Tomlinson as authors.
5. The Son Tay fiasco brings to mind *Team America*, a 2004 parody of the U.S military's global operations. Produced by the creators of *South Park*, the film portrays U.S. missions as hapless affairs, most of which end in comical failure leaving conditions worse than they found them.
6. See the *Politico.com* story on the exchange: https://www.politico.com/story/2015/07/trump-attacks-mccain-i-like-people-who-werent-captured-120317. See also Felicia Sonnez, "Donald Trump on John McCain in 1999: 'Does being captured make you a hero?'," *Washington Post*, August 7, 2018.
7. Howes, *Voices of the Vietnam POWs*, 234. Scott Blakey, *Prisoner at War: The Survival of Commander Richard A. Stratton* (Garden City, NY: Anchor Press, 1978), points to a class argument (205).
8. Hubbell, *P.O.W.*, 75.
9. The preoccupation of Americans with communist brainwashing is evinced in its unlikely appearances. Carol McEldowney in *Hanoi Journal, 1967* (Amherst: University of Massachusetts Press, 2007) recounts going to Hanoi in 1967 with a delegation of U.S. peace activists. She saw and heard things that brought into question Washington's version of events. In the journal she kept, she wondered on October 5 how she would be able to report on the trip "without seeming brainwashed" (58).
10. The details of Knutson's capture are from Hubbell, *P.O.W.*, 91. Many POWs remember the anger of villagers directed at them but, contra Knutson's story that villagers were "egged on by an officer," others say that regular Army personnel protected them from the villagers. Bob Fant recalls (interview with Tom Wilber, March 15, 2015) a soldier

standing with a rifle at "port arms" position defending Fant from approaching locals. Some, like Frank Anton in *Why Didn't You Get Me Out? Betrayal in the Viet Cong Death Camps; The Truth about Heroes, Traitors, and Those Left Behind* (Arlington, TX: Summit, 1997), recall villagers' acts of kindness (32).

11. The battering of Knutson as described by Hubbell, in addition to the injuries he suffered in the shootdown and the absence of any medical attention, make remarkable Hubbell's report that by October 29 he was already recovering. Hubbell, *P.O.W.*, 116.
12. Ibid., 98.
13. Hubbell's description of Alvarez's expectations that he will be hung by his ankles, skinned, castrated, and decapitated are intriguing. There is no footnote for that description, which leads us to wonder: is this Hubbell's imagination at work or Alvarez's? In either case, it reads as a deprecation of the Asian Other, born of Occidentalist prejudices. Used uncritically, as it is by Hubbell, the words function as stage-setting for the prisoner-at-war narrative the Hubbell is constructing. Hubbell, *P.O.W.*, 7.
14. Interview by Tom Wilber, April 4, 2017, in Hanoi with Nguyen Minh Y, retired army officer, who, as a junior officer, worked in the detention camps from the internment of Alvarez at Hoa Lo until calling off the names one-by-one at the release of the last group of prisoners at Gia Lam airport. See also Malcolm W. Browne, "Thousands Watch 67 Prisoners Depart," *New York Times*, March 30, 1973.
15. Hubbell, *P.O.W.*, 118.
16. Note the careful wording in Knutson's solely witnessed Silver Star citation: "For gallantry and intrepidity in action against the enemy in North Vietnam on 17 October 1965. Shortly after parachuting onto enemy soil, he was surrounded by village militia armed with rifles. In the face of great personal risk, he elected to fight rather than surrender. Defending himself with his *service revolver, he shot at his rifle-armed adversaries, inflicting two casualties* prior to being overwhelmed by their superior numbers. By his daring actions, extraordinary courage, and aggressiveness in the face of the enemy, he reflected great credit upon himself and upheld the highest traditions of the Naval Service and the United States Armed Forces" (italics added). The revolver he carried was not a true "service revolver" in the sense of a weapon but was issued and loaded as a search-and-rescue signaling device.
17. Knutson admitted the illegality of the shooting in an interview with Darrel Ehrlick for the Billings, Montana *Gazette*, November 11, 2015,
18. As a result of Tom Wilber's inquiries seeking documentation in the detention camp system, retired Hoa Lo staff helped locate in 2018 several caches of more than 100 copies of antiwar newspapers, including GI-published titles such as *GI Press Service*, *The Bond*, *EPF Newsletter*,

Resistance, War Bulletin, The Second Front Review, YLO, Liberation News Service, Fort Lewis Free Press, which are in the process of being catalogued into the collection at Hoa Lo Prison Museum. These are actual papers made available to prisoners, some dated as early as 1968.

19. Hubbell, *P.O.W.*, 262.
20. Rochester and Kiley, *Honor Bound*, write, "Prisoners were aware of the Russell inquest and . . . the death of Norman Morrison" (193).
21. McEldowney, *Hanoi Journal*, 97.
22. Rochester and Kiley, *Honor Bound*, 442.
23. Mary Hershberger, *Traveling to Vietnam*, 23.
24. James Clinton, *The Loyal Opposition: Americans in North Vietnam, 1965–1972* (Niwot, CO: University of Colorado Press), 10. A delegation from the United States, Women's Strike for Peace, also went in the summer of 1965 but did not meet POWs at that time.
25. Staughton Lynd and Tom Hayden, *The Other Side* (New York: New American Library, 1966), 100.
26. Clinton, *The Loyal Opposition*, 18. Aptheker does not name the POWs they met.
27. Hubbell, *P.O.W.*, 438.
28. Nguyen Minh Y, one of the few English-speaking NLF officers in the prisons, and assigned to Hoa Lo from the arrival of the first prisoner, Alvarez, said that many pilots gave interviews and recorded statements purely as a way to let their families know that they were well treated. Mr. Y reflected that it was Nixon's fabrication of the POW treatment issue that caused POWs to be concerned about their families suffering from unfounded worries. Interview with Tom Wilber, Hanoi, May 6, 2019.
29. Hubbell, *P.O.W.*, 549. There are no footnotes in the book so we can't tell how he arrived at those numbers.
30. George Coker as quoted in Barbara Powers Wyatt, ed., *We Came Home* (Toluca Lake, CA: P.O.W. Publications, 1977).
31. Tom Wilber interview with Nguyen Minh Y, Hanoi, April 4, 2017. We discussed in detail the case of David Wesley Hoffman, his interview with George Wald where he detailed his medical care for his compound broken arm, his lengthy monologue filmed by interviewers from Medical Aid to Indochina in May 1972 criticizing the massive escalation of bombing, his participation among interviewees of Jane Fonda, his report of his good medical care immediately on his return, and the sudden reversal in his narrative upon his return to the United States. Mr. Y attributes this to opportunism and understands that returning prisoners needed to say what they needed to in order to reintegrate into their culture and careers and get on with their lives. In an interview on May 16, 2016, in his home in Haiphong, Tran Trong Duyet told Tom Wilber that the importance to him personally of ensuring that

prisoners were treated fairly was his respect for family. He relayed the grief felt in his family when the French came to his home in 1951, took his older brother outside and beheaded him. He did not want families of prisoners to worry or grieve.

32. McEldowney, *Hanoi Journal, 1967*, 100.
33. The HBO documentary *Jane Fonda in Five Acts* has video of Hoffman describing this after his release: "If Miss Fonda thinks for a moment that any of the people that she saw were able to speak freely she's got another thing coming . . . I think coerced is a very mild word. I've used the word 'torture' initially." Note that Hoffman is not claiming he was tortured, but that he used the word. Hoffman was shot down late in 1971 and torture was reported to have ended in 1969. Fonda: "I think they're lying and I think they're not only going to have to live with the fact that they were carrying out acts of murder for the rest of their lives, they're also going to have to live with the fact that they are lying." See also Vanderbilt Television News Archive, https://tvnews.vanderbilt.edu, *CBS News*, Friday, April 13, 1973, record number 228071.
34. Hubbell, *P.O.W.*, 576.
35. James Daly and Lee Bergman, *Black Prisoner of War: A Black Conscientious Objector's Vietnam Memoir* (Lawrence: University Press of Kansas, 2000), 190.
36. Anton, *Why Didn't You Get Me Out?*, 32.
37. Howes, *Voices of the Vietnam POWs*, 218. Our use of "Other" is in the anthropological sense of the term.
38. "Now, Mayhew judges the US aggression unjust. He says, moreover, on the radio and in the papers that his compatriots condemn it more and more vigorously. Two American women that he met in Hanoi confirmed this information, and moreover, he saw films on Moratorium Day in November 1969, in the United States." Theo Ronco, "How American Pilots Captured in North Vietnam Live," *L'Humanité*, November 5, 1970.
39. British journalist Felix Greene was one of the first Western reporters to cover the war from Vietnam. His *Vietnam! Vietnam!* (Palo Alto, CA: Fulton Publshing Company, 1966) was made available to prisoners by the wardens. Carrigan's reference to stateside prisons was his recognition that, after the war, he could be court-martialed and jailed for what he had said. See McEldowney, *Hanoi Journal, 1967*, 96–97.

2. Profiles of Dissent: Senior Officers

1. The minimum accepted IQ score for admission to Mensa International, sometimes called the "genius society," is 132 on the Stanford-Binet test and 148 on the Cattell test.
2. Miller responds to interviewers regarding his military career: Edison W. Miller Collection, (AFC/2001/001/33509), Veterans History Project,

American Folklife Center, Library of Congress, https://memory.loc.gov/diglib/vhp/story/loc.natlib.afc2001001.33509/

3. Wilber tells all of this in an oral history video held in the Walter Eugene Wilber Collection, AFC/2001/001/69160, Veterans History Project, American Folklife Center, Library of Congress, http://memory.loc.gov/diglib/vhp/bib/69160.
4. Wilber and Miller established their combat bona fides in the air over Korea, whereas neither Stockdale nor Denton flew in combat in Korea—somehow, they missed that. That they were hostile to Wilber and Miller is fact; efforts to account for that hostility should consider that the accomplishments of the latter amplified their own feelings of inadequacy.
5. In the movie *The Fog of War: Eleven Lessons from the life of Robert S. McNamara*, McNamara provides this explanation for the nature of his departure: "And I said to a very close and dear friend of mine, Kay Graham, the former publisher of the *Washington Post*, 'Even to this day, Kay, I don't know whether I quit or was fired.' She said, 'You're out of your mind. Of course you were fired.'"
6. Michael O'Connor, *MiG Killers of Yankee Station* (Friendship, WI: New Past Press, 2003), gives details how Wilber, after being ordered by air control to shoot down a MiG and then locked on to his target, was suddenly ordered by the air war commander to abort and depart the area. Wilber turned away from the MiG to return to the ship and another, undetected MiG shot him from behind (126–28).
7. In an interview on May 5, 2019, with Tom Wilber, former Hoa Lo prison supervisor Nguyen Minh Y used the term "single room" to describe how Wilber was housed at Hoa Lo. When questioned about the term "solitary confinement," Mr. Y was not in full agreement with the use of the term. He described the solo nature of solitary as having to do with the number of people in a room and that the prisoner was allowed to leave the room daily for bathing and meals and getting some exercise. To Mr. Y, coupling "solitary" with "confinement" would connote never being able to leave the cell, which, he said, was not the case. He stated that, in their opinion, having the privacy of living in a small room was more favorable and that it afforded more protection. He said that although the North Vietnamese Army eventually gained control of the Hoa Lo facility from the Hanoi police, they initially leased only "four or five rooms" for captured pilots, Lieutenant Junior Grade Alvarez being the first in August 1964. It was only when those first few rooms were full of single occupants and newly captured pilots arrived that the staff began to double up the prisoners. When the French controlled the prison, the population density of Hoa Lo was about four to five times the U.S. prisoner population density.
8. Ralph Gaither, *With God in a P.O.W. Camp* (Nashville, TN: Boardman Press, 1973), 26.

9. Quoting POW memoirs, Hubbell and Rochester and Kiley use a mix of "cell" and "room" when referring to the holding facilities. Wilber's later interview was with Mike Wallace for CBS's *60 Minutes* after his release and return on April 1, 1973.
10. In a May 6, 2019, interview with Tom Wilber, Nguyen Minh Y pointed out that the DRV did not build prison facilities for the pilots. The army gradually took over Hoa Lo prisons from the Hanoi police. Other than Hoa Lo, the facilities were not designed as prisons but were repurposed as such. For example, the Zoo site was selected because it was a compound for filmmaking built by the French that included kitchen and worker housing and a secure wall and gates to protect from theft.
11. Gerald Coffee, *Beyond Survival: Building on Hard Times—A POW's Inspiring Story* (Aiea, Hawaii: Coffee Enterprises, Inc., 2013), 240.
12. A copy of the transcript is in the author's possession. The portions used here have been edited for length and clarity only.
13. Tom Wilber received a greeting from his father on his fifteenth birthday, June 19, 1970:

    Announcer: Here is the birthday message to his son from Walter Eugene Wilber, Commander, U.S. Navy, American pilot captured in the Democratic Republic of Vietnam.

    Wilber:

    To: Mister Thomas Eugene Wilber, 3212 Edinburgh Drive, Virginia Beach, Virginia 23452, U.S.A.

    From: Walter Eugene Wilber, Commander, U.S. Navy, Camp of Detention for US Pilots Captured in the Democratic Republic of Vietnam.

    Dear Thomas, happy birthday! I hope June nineteenth will be a very happy day, Tom. I wish you the very best of health and happiness, and wish, too, that I could be with you on your fifteenth birthday, however, I send you lots of love from my heart. . . .

    Have a happy summer vacation but remember to be safe. I know you, and your brothers and sister, are being helpful to Mommy.

    You and Bruce are old enough now to work for peace. . . . I am fine. Happy birthday, Tom.

    Love, Dad.

    Announcer: That was Walter Eugene Wilber, Commander, U.S. Navy, American pilot captured in the Democratic Republic of Vietnam, addressing his son on the occasion of his birthday.
14. Special to the *New York Times*, "Laird Discounts P.O.W. Interviews," December 29, 1970; Christopher Lydon, "Camp Termed Showplace," *New York Times*, December 29, 1970.
15. Hubbell, *P.O.W.*, 558. Hubbell provides no citation for the quote.
16. Ibid., 558.

17. Ibid., 558–59.
18. See discussion on orders to take torture, not to make statements or write letters, in Stephen A. Rowan, *They Wouldn't Let Us Die: The Prisoners of War Tell Their Story* (Middle Village, NY: Jonathan David Publishers, 1973), 165.
19. Hubbell, *P.O.W.*, 559, 561.
20. Bill Zimmerman, *Troublemaker: A Memoir from the Front Lines of the Sixties* (New York: Doubleday, 2011), 249; NVN: *Ramsey Clark Visit to POW Camp*, Library of Congress POW/MIA Database and Documents, CIA Files, Reel 408, vol. 28, 35, http://lcweb2.loc.gov/frd/pwmia/405/100645.pdf; NVN: *Jane Fonda Meets With U.S. POWs in Hanoi*," Library of Congress POW/MIA Database and Documents, CIA Files, Reel 408, vol. 25, 12–14, http://lcweb2.loc.gov/frd/pwmia/408/119397.pdf.
21. Tom Wilber interview with Lynn Guenther, March 8, 2018.
22. *North Vietnam, pre-1975: Letter to Walter Cronkite of CBS News Regarding Bombings in the Democratic Republic of Vietnam*, Library of Congress POW/MIA Database and Documents, Senate Committee Source Data Files, Reel PDS88, http://lcweb2.loc.gov/frd/pwmia/PDS88/127954.pdf.
23. William Beecher, "Laird Discusses Released P.O.W.s," *New York Times*, September 12, 1972.
24. Copies of the tape and transcript are in the author's possession. Portions of the transcript have been edited for length and clarity.
25. Dick Baumbach "Wilber Silent on Charges," *Elmira Star-Gazette*, June 27, 1973; Hubbell, *P.O.W.*, 601–3.
26. Associated Press, "Navy Censures Former POWs for Misconduct," *Cornell Daily Sun*, September 28, 1973.

3. Profiles in Dissent: "The Peace Committee" of Enlisted POWs

1. Howes, *Voices of the Vietnam POWs*, 108. "The Eight," known as the Peace Committee, are also referred to as "PCs" by some historians.
2. Wilber and Miller were not part of the Peace Committee, there being no contact before the Paris Peace Accords were signed.
3. Guy as told to Grant, *Survivors*, 253–55.
4. Ibid., 211–12, 334.
5. Rochester and Kiley, *Honor Bound*, 268–69.
6. Daly. *Black Prisoner of War*, 50, 63.
7. Captain Floyd "Hal" Kushner was an Army doctor who was captured November 1968. Initially held and moved with ten South Vietnamese army prisoners, he arrived at the camp with other Americans in January 1968. There, and until the group reached Hanoi in spring 1971, he treated fellow captives using such rudimentary resources available. While he was held in Hanoi, his wife, Valerie, endorsed the antiwar

candidacy of George McGovern for president, a story used by filmmakers Ken Burns and Lynn Novick in their 2017 film *The Vietnam War*, to support a female-betrayal narrative for the loss of the war.

8. Daly refers to the site as Plantation Gardens. The "gardens" referred to a small mound in the courtyard planted with flowers. After three years of jungle-living and forty-eight days of rough travel on the Ho Chi Minh Trail, the basketball hoops in the courtyard and electric lighting in their rooms—not cells—the change was, wrote Daly in *Black Prisoner of War,* "like going from Hell to Heaven" (174). Elliott Gruner, *Prisoners of Culture: Representing the Vietnam POW* (New Brunswick, NJ: Rutgers University Press, 1993), contrasts Daly's description of Hanoi conditions with the majority of pilot memoirs that used "hell" to describe their experience.
9. Daly, *Black Prisoner of War*, 184. Although the Kushner group knew about the PCs and *New Life*, they were kept separated by the guards for several months. Dr. Kushner and Frank Anton, a warrant officer helicopter pilot, were moved to a holding area for officers.
10. Ibid., 186–87. In his memoir, Daly doubts that most of the letters ever made it out of Vietnam, speculating that they were "filed" for later use by prison authorities.
11. Ibid., 190, 194. Emphasis added.
12. Gruner, in *Prisoners of Culture*, refers to an "unsettling consecration of experience" in the memoirs "written by white male pilots" that then accounts for their supposed emergence from imprisonment as better men (149).
13. Daly, *Black Prisoner of War,* 10–12, 39. The Jehovah's Witnesses grew out of an antiwar group during the First World War known as The Bible Students, or the Russellites, named for Charles Russell, a Bible fundamentalist. In June 1918, the group's president and six other leaders were convicted under the federal Espionage Act and sentenced to twenty years in prison. In 1931, the group was renamed the Jehovah's Witnesses. Zoe Knox, "'A Greater Danger than a Division of the German Army': Bible Students and Opposition to War in World War I America," *Peace & Change: A Journal of Peace Research* (April 2019), 207–43.
14. Daly, *Black Prisoner of War,* 10–12.
15. Ibid., 1–4. The quoted words are Daly's, 4.
16. Ibid., 14.
17. Ibid., 40.
18. Ibid., 209.
19. The details of Chenoweth's background are compiled from Michael E. Ruane, "Traitors or Patriots? Eight Vietnam POWs Were Charged with Collaborating with the Enemy," *Washington Post*, September 22, 2017; and Chenoweth's correspondence with Lembcke, March 7, 2019.

20. See Rochester and Kiley, *Honor Bound,* 446, 456, for details on Portholes.
21. "Work" in the prison camps like Portholes meant foraging for food and water. The prisoners were sometimes lightly watched by guards and allowed some contact with one another and local villagers.
22. The details on Riate are taken from Jerry Lembcke, correspondence with Bob Chenoweth, March 14, 2019.
23. Riate's song in Vietnamese can be accessed on YouTube: https://www.youtube.com/watch?v=OLtc3YevH8o&feature=youtube.
24. Daly, *Black Prisoner of War,* 208. Daly continued, "As the raid went on, I was suddenly filled for the first time in my life with real hate. I hated every plane and every pilot who flew them. And I felt so sick at what was happening outside that window—so sick and ashamed and sad at what my country was doing—that I started to cry."
25. Daly, *Black Prisoner of War,* 208.
26. Xuan Oanh also asked the Peace Committee if they wanted to leave Hanoi interspersed with the other prisoners or as group. The Eight, aware of the potential threat, chose safety in numbers, remaining in their cell block and choosing to be released together for the March 16, 1973, flights from Hanoi. Tom Wilber interview with Bob Chenoweth, Hanoi, July 4, 2019.
27. Once stateside, Guy walked back his threats of violence against the dissidents. Howes, *Voices of the Vietnam POWs*, 109.
28. Zalin Grant, *Survivors* (New York: W. W. Norton, 1975), 210.
29. Ibid., 272.
30. Ibid., 275.
31. John Young also recalled making a tape for Christmas broadcast in 1968 that would let his family know he was alive, but it was never used. In July 1969, he was asked by an official to make an antiwar statement for broadcast, but he refused. Ibid., 274–76.
32. Ibid., 280–81.
33. Hubbell, *P.O.W.,* 505, 508–509. Hubbell attributes those reports of abuse to Dennis Thompson and Edward Leonard but does not say where, when, and to whom they claimed such.
34. Grant, *Survivors*, 276.
35. Ibid., 276.
36. Hubbell, *P.O.W.,* 567–68. Since *P.O.W.* is not footnoted, we don't know the source of the memo Hubbell refers to. By that time, however, in-service GIs and Vietnam veterans were making similar statements quite widely so there is no reason to question the general validity of his claim against Branch.
37. Grant, *Survivors*, 245–46.
38. Daly, *Black Prisoner of War,* 135.
39. Rochester and Kiley, *Honor Bound,* 457.

40. Hubbell, *P.O.W.*, 533. Again, without footnotes, we cannot know his source for this. It is noteworthy that the anecdote does not appear in Daly or Grant, which are the two best primary sources on Kavanaugh. Rochester and Kiley, who wrote their book twelve years after Hubbell, and rely on him for other details, do not repeat the story.
41. Grant, *Survivors*, 280–82.
42. Daly, *Black Prisoner of War*, 203–4.
43. Grant, *Survivors*, 317.
44. Ibid., 338.
45. Ibid., 338.
46. The Pentagon's press release is reproduced in ibid., 340.
47. Rochester and Kiley's remark is in *Honor Bound*'s page 563 footnote that has within it a set of citations to a chief of staff memo, and articles from the *New York Times* and *Washington Star & News;* Lieutenant Colonel Gary D. Solis, *U.S. Marine Corps, Marines and the Military Law in Vietnam: Trial by Fire* (Washington, DC: U.S. Marine Corps History and Museums Division, Headquarters, 1989), 218–19; Grant, *Survivors*, 334–43.
48. Daly, *Black Prisoner of War*, 265.
49. Silber was called before the House Un-American Activities Committee (HUAC) in the 1950s. He became famous for his edited collections of folk and political music, *Sing Out*, and *Folksinger's Workbook*. During the Vietnam War, he edited the widely read *National Guardian*, a newspaper that featured the work of Australian journalist Wilfred Burchett. Dane was a folk, blues, and jazz singer who performed with Muddy Waters, Memphis Slim, and Pete Seeger. According to Wikipedia, she was the first American musician to tour in post-revolutionary Cuba.
50. Chenoweth correspondence with Lembcke, March 5, 2019. According to Chenoweth, Riate was assassinated in 1984. He was thirty-nine.
51. In correspondence with Tom Wilber, July 16, 2018, Chenoweth recalls being told of Riate's assassination by his brother. Riate also worked with Indochina Peace Campaign (IPC), lending his words to an organizing poster that urged rejection of the SROs' claims of torture and called out the brutal treatment of prisoners by the U.S and Saigon forces. See Series III.ii, Folder 4 of the IPC files in Special Collections at the University of Massachusetts Library, Boston.

4. *The Manchurian Candidate* Stalks the Homeland: Hollywood Scripts the POW Narrative

1. Epigraph: Rochester and Kiley, *Honor Bound*, 562.
2. Among the studies debunking brainwashing was Robert Jay Lifton's 1961 *Thought Reform and The Psychology of Totalism: A Study of Brainwashing in China* (New York: W. W. Norton, 1961).
3. Remarkably, in early pages of *Honor Bound*, Rochester and Kiley cited a

reason for the Pentagon to drop the charges against the PCs, which was that they had acted with honor.

4. Risner, *The Passing of the Night: My Seven Years as a Prisoner of the North Vietnamese* (New York: Random House, 1973), 91–92.

5. There might be some historical interpolation in Howes's assessment of how seminal *The Manchurian Candidate* was in the myth-making that grew out of the Korean War POW experience and how influential it was on the pilots destined for Vietnam. In a 2003 recall of the film's history, critic Roger Ebert points out that it had only a two-year run, 1962 to 1964, and was not re-released until 1988. Its initial popularity, moreover, was due to its allegorical reference to the Kennedy assassination in 1963, more than what it said about the war in Korea. Ebert recalled being told by John Frankenheimer, the film's director, that Frank Sinatra, who starred in the film, "purchased the rights [to the film] and kept it out of release from 1964 to 1988 . . . [out of] remorse after Kennedy's death."

6. The 1954 *The Bamboo Prison*, out just a year after fighting ended in summer 1953, appears now as a shakedown cruise by a Hollywood unsure about what to do with the Korean War POW story. On the one hand, the film has throwback appeal to a Second World War film like *Stalag 17*, including the slapstick humor that would be the hallmark of the 1960s television series *Hogan's Heroes*. On the other hand, *Bamboo Prison* pioneers the themes of brainwashing and collaboration with the enemy that will lead to *The Manchurian Candidate* and dominate the POW genre from then on.

7. Elaine May, *Fortress America: How We Embraced Fear and Abandoned Democracy* (New York: Basic Books, 2017).

8. The interest of fliers and flight crews would have been heightened by depictions of pilots shot down and captured in Korea. But how many of those were there in the war? For his *Remembered Prisoners of a Forgotten War*, historian Lewis H. Carlson interviewed Robert Coury, who told of being shot down and captured in June of 1953. Coury recalled meeting three more captive pilots at an interrogation center. In the 1954 film *Bridges at Toko-Ri*, Navy Lieutenant Harry Brubaker, played by William Holden, is shot down but killed by enemy troops before capture. Based on a book by James Michener, with Grace Kelly and Mickey Rooney in supporting roles, the film was hugely popular. Lewis, Carlson, *Remembered Prisoners of a Forgotten War* (New York: St. Martin's Press, 2002), 35.

9. In his speech "The Ultimate Weapon" on communist mind control, former Major William Mayer, who was the primary source for Kinkead, said, "We thought we knew about the burning bamboo splints under the fingernails used by all Orientals." All the ordinary U.S. soldiers knew this, he said. But he went on to say that 95 percent of the men

saw no such physical abuse. https://www.usa-anti-communist.com/pdf1/Mayer_Brainwashing_Ultimate_Weapon/Brainwashing_The_Ultimate_Weapon-Major_William_E_Mayer-Oct4_1956.pdf

10. Paul Hanebrink, *A Specter Haunting Europe: The Myth of Judeo-Bolshevism* (Cambridge, MA: Belknap Press, 2019), writes the history of the racializing of anti-communism.
11. Vietnam War and Second World War captivities contrast in other ways. Germans, for example, followed the Geneva guidelines for treatment. Japan did not follow those guidelines and did murder some POWs. Notably, thousands of prisoners taken by Japan are thought to have died on ships sunk by U.S. submarines and air attacks.
12. The scenes of capture resemble those in the Second World War classic *Bridge On the River Kwai*, replete with the captives whistling a marching cadence, whereas the Viet Minh are portrayed as buffoonish as the Germans in *Stalag 17*.
13. In the film figure cut by Raspeguy, we can recognize a composite profile of the real-life Vietnam POWs Edison Miller and Gene Wilber—modest class background, affinity with the racial/ethnic Other, and aversity with military authority.
14. Hershberger, *Traveling to Vietnam*, 143–44.
15. George Smith's story is told by Howes in *Voices of the Vietnam POWs*.
16. Chapter 9 of Lembcke's *The Spitting Image: Myth, Memory, and the Legacy of Vietnam* (New York: New York University Press, 1998) uses archives from the Margret Herrick Library and director Waldo Salt's papers at the UCLA library to document the making of *Coming Home*.
17. A film proposed by Old Westbury College professor Steve Talbot in April 1973, titled *The Man in the Sky Is a Killer*, is illustrative of New Left attitudes of the POWs. The script describes POWs as "almost all loyal officers; they are career men, volunteers . . . who recall an America before all this trouble started. Like Cold War Rip Van Winkles, they praise the December [1972] bombing, and attack the people who were downing or bad-mouthing our government policies." Parenthetically, the proposal included a nod to a place in the film for the Peace Committee. The title of the film is said to be a Vietnamese saying.
18. The association of the American Left with global communism was taken seriously in prowar conservative circles during the war. See Hon. John G. Schmitz, *The Viet Cong Front in the United States*, read into the *Congressional Record*, April 21, 1971, as *The Second Front of the Vietnam War: Communist Subversion of the Peace Movement*. Schmitz was a congressman from Orange County, California, and a member of the John Birch Society.
19. See Hershberger, *Traveling to Vietnam*, for a full account of American antiwar delegations to Hanoi.
20. McEldowney's biographical details are taken from Suzanne Kelley

McCormack's Introduction to McEldowney's *Hanoi Journal, 1967*, which McCormack edited. The Rothstein details are from her interview with the PBS/WGBH series for *Peoples' Century*, "Young Blood: 1950–1975," https://www.pbs.org/wgbh/peoplescentury/episodes/youngblood /rothsteintranscript.html.

21. The other members of the Bratislava Hayden delegation were ERAP leader Rennie Davis, filmmaker Norman Fruchter, and Robert Allen and John "Jock" Brown, who joined the group in Bratislava.
22. McEldowney, *Hanoi Journal, 1967*, 58.
23. Ibid., 94–95.
24. After the meeting with Carrigan, Rothstein wrote: "I felt he was baiting and bullshitting us—because he wants to please the Vietnamese to get good treatment."
25. See Franklin's *M.I.A. or Mythmaking in America: How and Why the Belief in Live POWs Possessed a Nation* (New York: Lawrence Hill Books, 1992).
26. *When Hell Was in Session* (1979) was the first Vietnam POW-themed film. Set in the Hanoi prisons that the prisoners called "The Zoo" and "Alcatraz," it was based on Rear Admiral Jeremiah Denton's memoir of the same title. In a pseudo-documentary form, the film followed the chronology of Denton's shootdown, capture, interrogations, brutal treatment, isolation, Christian commitment, and release, with scenes interspersed of his family back home enduring his absence. The only characters in the film are Air Force and Navy pilots—officers. There is no hint of enlisted men among them or hint of divisions within the officer ranks.
27. Film critic Tony Williams in Malo and Williams, *Vietnam War Films*, reprised *Uncommon Valor*, attributing its production to, "The alliance of producer John Milius, an avowed right-wing militaristic Hollywood movie brat, and Ted Kotcheff, the director of *First Blood* that inaugurated the *Rambo* series." Jean-Jacques Malo and Tony Williams, *Vietnam War Films* (Jefferson, NC: McFarland & Company, 1994), 450.
28. See David Sirota, *Back to Our Future: How the 1980s Explains the World We Live In Now—Our Culture, Our Politics, Our Everything* (New York: Ballantine Books, 2011), for 1980s popular revisionist influences on public memory of the Vietnam War.

5. Damaged, Duped, and Left Behind

1. For a history of the coffeehouses see David L. Parsons in *Dangerous Grounds: Antiwar Coffeehouses and Military Dissent* (Chapel Hill: University of North Carolina Press, 2017); for GI Press history, see James Lewes, *Protest and Survive: Underground GI Newspapers during the Vietnam War* (New York: Praeger, 2003).
2. An Article 15 violation of the Uniform Code of Military Justice is heard in

an administrative proceeding for offenses akin to a misdemeanor in civilian law. An Article 15 guilty verdict is meted as nonjudicial punishment.

3. The military constraints on civilian reporters were tighter than sometimes believed today. It was not easy for reporters to get outside of major cities and military installations. When they did, they were sometimes "given a story" by a field unit's "public affairs liaison" and put on a plane back to Saigon. See David Cortright, *Soldiers in Revolt* (New York: Anchor Books, 1975), 269, for references to the Army's inquiries into dissent.
4. See Andrew Hunt, *The Turning: A History of Vietnam Veterans Against the War* (New York: New York University Press, 2001).
5. The University of Northern Colorado chapter of VVAW, of which Jerry Lembcke was a member, was banned from a Veterans Day parade in the early 1970s. Working around the ban, it followed behind the parade stepping to a solemn "death march" cadence.
6. For an analysis of Kerry's speech and responses to it, see John Kerry, *The New Soldier: Vietnam Veterans Against the War,* edited by David Thorne and George Butler (New York: Collier, 1971).
7. John R. Coyne, *The Impudent Snobs: Agnew vs. The Intellectual Establishment* (Arlington, VA: Arlington House Press, 1972).
8. Erving Goffman, *Stigma* (New York: Touchstone, 1964/1986).
9. See Daniel Ellsberg, *Secrets: A Memoir of Vietnam and the Pentagon Papers* (New York: Penguin, 2003).
10. The *Times* article was Jon Nordheimer, "Postwar Shock Besets Ex-G.I.s," *New York Times,* August 21, 1972; Peter Bourne, *Men, Stress, and Vietnam* (New York: Little, Brown, 1970).
11. The blackout of news about shot-down U.S. pilots may never have been as great as some Americans believe today. In Hubbell's *P.O.W.,* 51–52, he records pilot Larry Guarino's arrival in Hoa Lo prison in June of 1965 and telling Bob Peel, who had been captured earlier, that he had read about Peel's capture and that "your name has been officially released as definitely captured." Peel was the eighth pilot captured and a month later Guarino was the tenth.
12. The SROs also played the "buyout" card, offering Wilber and Miller the chance for "reinstatement" as commanding officers in the chain of command they had configured. See Rochester and Kiley, *Honor Bound,* 553.
13. The origins of the "weakness" theory in "official" accounts of Korean War POWs is in Albert D. Biderman, *March to Calumny: The Story of American POWs in the Korean War* (New York: Macmillan, 1963), 166–67.
14. Hubbell attributes the Kushners' motivations to "naivete, weakness, and mental illness," Hubbell, *P.O.W.,* 109. Rochester and Kiley add "lacked strength and intelligence and discipline" to the list, Rochester and Kiley, *Honor Bound,* 565.
15. See Seymour Hersh's coverage of the Wallace interview in "P.O.W. Who Made Antiwar Statements in Hanoi Recalls 'Pressure of Conscience,' " *New York Times,* April 2, 1973.

16. The *New York Times*, "Ex-P.O.W.s Cheer Nixon," May 23, 1970, made no mention of the dissidents, nor did its "400 Ex-P.O.W.s are Given $400,000 Dallas Reception," June 2, 1970. Tom Wilber, Gene's son, is a source for the behind-the-scenes shenanigans against the family.
17. Later determinations made it clear that the Code of Conduct was open to interpretation and was not law. Being unrelated to the Uniform Code of Military Justice, no legal determination can be made based on a dispute over whether a servicemember was following the Code of Conduct.
18. Rochester and Kiley, *Honor Bound*, 102, attribute the introduction of tap code as a means of communication to Air Force captain Carlyle Harris in the summer of 1965. Harris recalled learning about the tap code during a coffee break (not as part of the curriculum) when he was in Air Force survival training. A footnote in Rochester and Kiley dates the system back to the First World War and notes it was also used in Korean POW camps. The five-by-five matrix coded twenty-five letters of the alphabet, *K* being dropped and substituted with the letter *C*. John Dramesi characterizes the tap code as useful for "pornography and entertainment" in this 2008 interview: https://www.chicagoreader.com/Bleader/archives/2008/10/10/john-dramesis-unflattering-memories-of-his-fellow-pow-john-mccain.
19. Solis, *Trial by Fire*, 220.
20. "The two were retired with administrative letters of censure and in lasting disgrace." Rochester and Kiley, *Honor Bound*, 568.
21. As with other antiwar veterans, the diagnostic framing given their views functioned politically and culturally more than medically. Press reports at the time portrayed POWs as healthy, and later medical reports confirmed that. POW memoirs written as late as the mid-1980s make no mention of PTSD or trauma.
22. These details are taken from Franklin, *M.I.A.*, 39, 48–49. Photographs of the tiger cages were published in the July 17, 1970, *Life*. See Don Luce, "The Tiger Cages of Vietnam," www.historiansagainstwar.org/resources/torture/luce.html.
23. Jim Stockdale and Sybil Stockdale, *In Love and War*, rev. 1984 edition (Annapolis, MD: Naval Institute Press, 1990), 297–98. In a 1969 interview, John Frankenheimer, director of *The Manchurian Candidate*, told Canadian film critic Gerald Pratley: "We consulted every book written about brainwashing, and I remember reading one called *In Every War But One*. . . . We believed that we lived in a society that was brainwashed. And I wanted to do something about it." The interview can be accessed at https://www.filmsocietywellington.net.nz/db/screeningdetail.php?id=2.
24. Stockdale and Stockdale, *In Love and War*, 299–300.
25. Franklin, *M.I.A.*, 196n35.
26. Ibid., 13–23, for details on the counting of POWs, MIAs, and BNR. Bernie Rupinski, Gene Wilber's radar intercept officer and backseater, is an example of a flier known to have died in the 1968 shootdown but whose body

was not recovered. In 2015, Tom Wilber located Rupinski's gravesite in Thanh Chuong District, Nghe An Province, Vietnam. Based on Tom's findings, DPAA (Defense POW/MIA Accounting Agency) reopened the case (REFNO 1209) but as of 2021, the remains are still unrecovered.

27. See L. C. McCollum, *History and Rhymes of the Lost Battalion*, (1919), and Arch Whitehouse, *Heroes and Legends of World War I* (New York: Doubleday, 1964) for more on the Legend of the Lost Command.
28. Franklin, *M.I.A.*, 54. Franklin's description is from "Exhibit to Stir Opinion on P.O.W.s Opens in Capitol," *New York Times*, June 5, 1970.
29. Ibid., 56–57.
30. Tom Wilber has POW/MIA bracelets for Walter Eugene Wilber and Bernard Francis Rupinski. One with Gene Wilber's name was returned to Tom in 2015 by a woman who had purchased it in the early 1970s and had worn in through the return of the POWs.
31. Franklin, *M.I.A.*, 5.
32. Jeremy Kuzmarov and John Marciano. *The Russians are Coming, Again* (New York: Monthly Review Press, 2018), reviews that history and updates its relevance for current American obsessions with Russian influence in domestic politics.
33. May's *Fortress America* delves into the social and cultural dimensions of the Cold War at home.
34. Perot's remarks on McCain as reposted here can be found at the *Newsweek* online site: https://www.newsweek.com/ross-perot-slams-mccain-86763.
35. Lembcke's *CNN's Tailwind Tale: Inside Vietnam's Last Great Myth* (Lantham, MD: Rowman & Littlefield, 2003) reveals the journalistic malfeasance and conspiratorial motif of the CNN report.

6. A Captive Nation: POWs as Grist for the American Myth

1. Paul Boyer,. *When Time Shall Be No More: Prophecy Belief in Modern American Culture* (Cambridge, MA: Harvard University Press, 1992), 68–69. In 1629 John White implored his fellow colonists to support the venture in New England as "a bulwark against the Kingdom of Antichrist," and preacher Cotton Mather interpreted the colonists' war against the Indians (King Philip's War of 1675–1676) as a manifestation of the Red Horse of the Apocalypse foretold in the Book of Revelation. A century later, with the colonies nearing their final struggle for independence, the Christian prophets regaled their British oppressors as the Antichrist and warned that the Stamp Act could be the Mark of the Beast in disguise.
2. Pauline Turner Strong. *Captive Selves, Captivating Others: The Politics and Poetics of Colonial American Captivity Narratives* (Boulder, CO: Westview Press, 1999), 48–51. The following paragraphs on the captivity narratives are indebted to Strong's book.
3. Ibid., 52–55. The absence of any corroborative testimony for Smith's reprieve adds to the question about its accuracy.

4. Bruce Cumings, *North Korea: Another Country* (New York: New Press, 2004), 13. Cumings acknowledges Anderson, *Imagined Communities,* for his phrasing. Another defacing and widely used term was a one-letter abbreviation for Vietnamese, "the V." Jay Jensen, in *Six Years in Hell: A Returned Vietnam POW Views Captivity, Country, and the Future* (Orcutt, CA: Publications of Worth, 1974), uses that expression almost exclusively. Writing about the food, for example, he said, "The 'V' told us not to worry about the worms" (124).
5. Eugene B. McDaniel, *Before Honor: One Man's Spiritual Journey into the Darkness of a Communist Prison (Before Honor Is Humility—Proverbs 18:12)* (New York: A. J. Holman, 1975), 50; Larry Guarino, *A P.O.W.'s Story: 2801 Days in Hanoi* (New York: Ivy Books, 1990), 139; Ralph Gaither. *With God in a P.O.W. Camp*, 132.
6. Gaither, *With God in a P.O.W. Camp*, 128.
7. See memoirs by Larry Chesley, *Seven Years in Hanoi: A POW Tells His Story* (Salt Lake City: Bookcraft, 1973), 109; and Jensen, *Six Years in Hell,* 104, on the letter campaign. The letter campaign is central to the "official story": see John G. Hubbell, *P.O.W.*, 519. Craig Howes, in *Voices of the Vietnam POWs,* mentions the Amnesty International claim, 106. John McCain with Mark Salter, in *Faith of My Fathers* (New York: Random House, 1999), give credence to the post-Ho logic for the change, 290; as does former POW Phillip Butler in *Three Lives of a Warrior* (Scotts Valley, CA: CreateSpace, 2010), 331–32.
8. Another twist in the cycle wherein the confession to the SRO that the prisoner had given the interrogators something came to represent proof of having been tortured, was that the confession compelled the subordinate POWs to degrade themselves as "failures" before their own superiors. Excellent on this point is Howes, *Voices of the Vietnam POWs*, chapter 3.
9. Howes, *Voices of the Vietnam POWs,* 70, draws on Elaine Scarry, *The Body in Pain: The Making and Unmaking of the World* (New York: Oxford University Press, 1985), 85, to make a social-psychological point that, just as the warrior gives consent for the infliction of bodily pain when he joins the battle, prisoners at war give consent for torture.
10. Stockdale and Stockdale, *In Love and War*, 332–38.
11. Risner, *The Passing of the Night,* 117–21.
12. "Insisting on torture," wrote Howes, in *Voices of the Vietnam POWs,* "was thus the POWs' way of fighting the Vietnam War and of guaranteeing they would return home proudly as victors in 'the battle of Hanoi'" (70).
13. There is no doubt that the hardcore SROs thought that signs of physical damage to their bodies could be later translated into propaganda statements about the cruelty of the Communists. But their recognition that the same bodily damage could be represented as a statement

about *them* is an even more important insight into the prisoner-*at*-war mentality.

14. What Risner dismissed as "propaganda" can be read today to say more about his misunderstandings of the war than for anything about the North Vietnamese. In his memoir, *The Passing of the Night*, for example, he wrote: "They kept preaching to us [over the radio] that North Viet Nam had been tricked [by the Geneva agreement that had ended the French occupation of Vietnam]. To hear them tell it, the Vietnamese people had been guaranteed that they would be reunited, and that all they really wanted were free elections. We knew that was a lie. The story of the Communist takeover in North Viet Nam was a three-day blood bath in which they murdered five hundred of the top men in the country to pave the way for Communism. Their 'freedom-loving people' jazz did not pull the wool over our eyes" (143).
15. One of Salisbury's reports, "A Visitor to Hanoi Inspects Damage Laid to U.S. Raids," was carried on page 1 of the December 25, 1966, edition of the *New York Times*. Salisbury described Hanoi and included a map issued by the U.S. State Department listing military targets and juxtaposed it with the civilian sites said by Vietnamese to have been hit in recent raids. Salisbury's report had the standard journalistic qualifiers in all the right places: "the North Vietnamese say . . ." "four persons were reported killed . . ." "Hanoi residents certainly believe they were bombed. . . ."—On a personal note, I (Lembcke) acquired Salisbury's book, *Behind the Lines—Hanoi*, published in 1967, just before going to Vietnam in early 1969. As Stockdale may have feared, the book was, for me, an eye-opener that raised serious questions about the truth of U.S. government accounts of the bombing and what the war was all about.
16. Stockdale, *In Love and War*, 245–49.
17. Howes, *Voices of the Vietnam POWs*, 95.
18. Dunn's recollection is in Barbara Powers Wyatt, ed., *We Came Home*, 1977. Air Force Captain Joseph Crecca had a B.S. degree in mechanical engineering and wrote in Wyatt, *We Came Home*, that, while a prisoner, he "had the opportunity to study Russian as well as to teach mathematics, physics, classical music, and automobile theory and mechanics" (8).

7. The Heritage of Conscience: From the American War in Vietnam to America Today

1. Seymour M. Hersh, "The Scene of the Crime: A Reporter's Journey to My Lai and the Secrets of the Past," *The New Yorker*, March 22, 2015.
2. Matthew Breems, "Courage to Resist Vietnam Series Podcast, Episode 19: Susan Schnall," August 22, 2019.
3. VVAW's Winter Soldier hearing is documented in Andrew Hunt's 1999 *The Turning* and the 1972 film *Winter Soldier*. The IVAW hearing is

described by Nan Levinson in *War Is Not a Game: The New Antiwar Soldiers and the Movement They Built* (New Brunswick, NJ: Rutgers University Press, 2014), 2014.

4. Jon Krakauer, *Where Men Win Glory: The Odyssey of Pat Tillman* (New York: Anchor Books, 2010).
5. According to Krakauer, Tillman's friend Reka Cseresnyes contacted Chomsky on Tillman's behalf and the professor was open to meeting him (263–64).
6. Ibid., 329–41. Destruction of his belongings was against regulations. Tillman's status as a celebrity figure *cum* Army poster boy spawned resentment among his peers. Jason Porter, the NCO responsible for Tillman's orientation to Ranger training, thought of him as a "prima-donna football star" who other NCOs sucked up to. Porter was in the platoon that killed Tillman. Knowing as we do that high ranking officers who resented Gene Wilber's dissent green-lighted an assault on him, it is not conspiracist to imagine that Tillman's resentful peers sensed a wink-and-a-nod from their higher-ups to take him out.
7. More details on the differences between duty in Iraq and Vietnam can be found in Jerry Lembcke, *PTSD: Diagnosis and Identity in Post-Empire America* (Lanham, MD: Lexington Books, 2013), 59–65.
8. Military and veteran dissent in the years of the war in Vietnam went through phases of official denial/suppression, criminalizing, and pathologizing, as we see in the cycle experienced by the dissenting POWs. Outright suppression and criminalizing of in-service and veteran dissent during the wars in Iraq and Afghanistan, by contrast, skipped the first two phases and picked up where the Vietnam trajectory left off, treating dissent as a mental and emotional disorder, a.k.a. PTSD.
9. Joseph B. Verrengia, "Some Iraq Vets Find Forgetting the Hardest Part About Killing," Associated Press (April 18, 2003).
10. The *Pittsburgh Tribune-Review* published "Stress of Battle Haunts Soldiers" in February 2005, and *USA Today* also published a feature in October 2007. A *New York Times* series that began on the front page of a mid-January 2008 Sunday edition ran for 5,600 words across three full pages, and continued for several more days. In *PTSD: Diagnosis and Identity in Post-Empire America*, Lembcke took a phenomenological dive into the *Times* series to ferret out its meaning.
11. Inspired by the Coffeehouse Movement of the Vietnam War years, antiwar veterans of Iraq opened coffeehouses near Fort Hood, Texas; Fort Lewis, Washington; and Fort Drum, New York. Ending the U.S. occupation of Iraq was part of their mission, but the new coffeehouses functioned more as service centers offering mental health counseling and help with personal and family issues.
12. Lolita C. Baldor and Robert Burns, "Reinstate? Reassign? Navy to

Decide Fate of Fired Captain," *The Associated Press*, April 19, 2020; Lolita C. Baldor and Robert Burns, "Navy Recommends Reinstatement of Fired Carrier Captain," *The Associated Press*, April 25, 2020; Eric Schmitt and Helene Cooper, "Navy to Pursue Wider Inquiry Into Actions Taken on Ship Hit by Coronavirus," *New York Times*, April 30, 2020.

13. https://www.usatoday.com/story/news/politics/2020/04/07/timeline-capt-crozier-firing-acting-navy-secretary-modly-resigning/2964617001/.
14. An April 16, 2020, "open letter" in *The Nation* magazine signed by nearly 100 former members of Students for a Democratic Society (SDS) invoked the experience of post–World War 1 Germany for its lessons for the then-upcoming Trump-Biden presidential election: https://www.thenation.com/article/activism/letter-new-left-biden/. Historian Van Gosse averred that the reelection of Donald Trump in 2020 threatened a martializing of American political culture: https://organizingupgrade.com/an-illiberal-democracy-if-trump-wins-again/.
15. Editorial, *New York Times*, May 18, 2020, https://www.nytimes.com/2020/05/18/opinion/trump-military.html.

SELECTED BIBLIOGRAPHY

Allen, Michael J. *Until the Last Man Comes Home: POWs, MIAs, and the Unending Vietnam War*, Chapel Hill: UNC Press, 2009.

Anton, Frank. *Why Didn't You Get Me Out? Betrayal in the Viet Cong Death Camps; The Truth About Heroes, Traitors, and Those Left Behind*. Arlington, TX: Summit, 1997.

Baumbach, Dick. "Wilber Silent on Charges," *Elmira Star-Gazette*, June 27, 1973.

Bates, Milton. *The Wars We Took to Vietnam: Cultural Conflict and Storytelling*. Berkeley: University of California Press, 1996.

Beecher, William. "Laird Discusses Released P.O.W.s." *New York Times*, September 28, 1972.

Biderman, Albert D. *March to Calumny: The Story of American POWs in the Korean War*. New York: Macmillan, 1963.

Blakey, Scott. *Prisoner at War: The Survival of Commander Richard A. Stratton*. Garden City, NY: Anchor Books, 1978.

Bourne, Peter. *Men, Stress, and Vietnam*. New York: Little, Brown, 1970.

Boyer, Paul. *When Time Shall Be No More: Prophecy Belief in Modern American Culture*. Cambridge, MA: Harvard University Press, 1992.

Breems, Matthew. "Courage to Resist Vietnam Series Podcast, Episode 19: Podcast Interview with Susan Schnall," August 22, 2019.

Browne, Malcolm. "Thousands Watch 67 Prisoners Depart." *New York Times*, March 30, 1973.

Burchett, Wilfred. *Mekong Upstream*. N/A: Red River Publishing House, 1957.

Chesley, Larry. *Seven Years in Hanoi: A POW Tells His Story*. Salt Lake City: Bookcraft, 1973.

Carlson, Lewis. *Remembered Prisoners of a Forgotten War*. New York: St. Martin's Press, 2002.

Chu Chí Thành, *Memories of the War*. Hanoi: Vietnam News Agency Publishing House, 2015.

Clements, Charles, M.D., *Witness to War: An American Doctor in El Salvador*. New York: Bantam Books, 1984.

Clinton, James W. *The Loyal Opposition: Americans in North Vietnam, 1965–1972*. Niwot, CO: University Press of Colorado, 1995.

Coffee, Gerald. *Beyond Survival: Building on Hard Times—A POW's Inspiring Story*. Aiea, Hawaii: Coffee Enterprises, Inc., 2013 (revised).

Condon, Richard. *The Manchurian Candidate*. New York: Orion, 2013 (reprint).

Cortright, David. *Soldiers in Revolt*. New York: Anchor Books, 1975.

Coyne, John R. *The Impudent Snobs: Agnew vs. The Intellectual Establishment*. Arlington, VA: Arlington House Press, 1972.

Cumings, Bruce. *North Korea: Another Country*. New York: New Press, 2004.

Daly, James A., and Lee Bergman. *Black Prisoner of War: A Conscientious Objector's Vietnam Memoir*. Lawrence: University Press of Kansas, 2000. First published in 1975 by Bobbs-Merrill, under the title *A Hero's Welcome*.

Denton, Jeremiah, and E. Brandt. *When Hell Was in Session*. Clover, SC: Commission Press, 1976.

Ehrlick, Darrel. "Interview with Rodney Knutson," *Gazette* (Billings, MT), November 11, 2015.

Ellsberg, Daniel. *Secrets: A Memoir of Vietnam and the Pentagon Papers*. New York: Penguin, 2003.

Franklin, H. Bruce. *M.I.A. or Mythmaking in America: How and Why the Belief in Live POWs Possessed a Nation*. New York: Lawrence Hill Books, 1992.

Gaither, Ralph. *With God in a P.O.W. Camp*. Nashville, TN: Broadman Press, 1973.

Goffman, Erving. *Stigma*. New York: Touchstone, 1964/1986.

Grant, Zalin. *Survivors*. New York: W. W. Norton, 1975.

Greene, Felix. *Vietnam! Vietnam!* Palo Alto, CA: Fulton Publishing Company, 1966.

Gruner, Elliott. *Prisoners of Culture: Representing the Vietnam POW*. New Brunswick, NJ: Rutgers University Press, 1993.

Guarino, Larry. *A P.O.W.'s Story: 2801 Days in Hanoi*. New York: Ivy Books, 1990.

Hanebrink, Paul. *A Specter Haunting Europe: The Myth of Judeo-Bolshevism*. Cambridge, MA: Belknap Press, 2019.

Hersh, Seymour M. "The Scene of the Crime: A Reporter's Journey to My Lai and the Secrets of the Past," *The New Yorker*, March 22, 2015.

———. "P.O.W. Who Made Antiwar Statements in Hanoi Recalls 'Pressure of Conscience,'" *New York Times*, April 2, 1973.

Hershberger, Mary. *Traveling to Vietnam: American Peace Activists and the War*. Syracuse, NY: Syracuse University Press, 1998.

Howes, Craig. *Voices of the Vietnam POWs: Witnesses to Their Fight*. New York: Oxford University Press, 1993.

Hubbell, John G. *P.O.W.: A Definitive History of the American Prisoner-of-War Experience in Vietnam, 1964–1973*. New York: Reader's Digest Press, 1976.

Hunt, Andrew. *The Turning: A History of Vietnam Veterans Against the War*. New York: New York University Press, 2001.

Jensen, Jay R. *Six Years in Hell: A Returned Vietnam POW Views Captivity, Country, and the Future*. Orcutt, CA: Publications of Worth, 1974.

Kerry, John. *The New Soldier: Vietnam Veterans against the War*. Edited by David Thorne and George Butler. New York: Collier, 1971.

Knox, Zoe. "A Greater Danger than a Division of the German Army: Bible Students and Opposition to War in World War I America," *Peace & Change* 44/2 (April 2019): 207–43.

Krakauer, Jon. *Where Men Win Glory: The Odyssey of Pat Tillman*. New York: Anchor Books, 2010.

Kuzmarov, Jeremy, and John Marciano. *The Russians are Coming, Again*. New York: Monthly Review Press, 2018.

Lembcke, Jerry. *PTSD: Diagnosis and Identity in Post-Empire America*. Lanham, MD: Lexington Books, 2013.

———. *CNN's Tailwind Tale: Inside Vietnam's Last Great Myth*. Lanham, MD: Rowman & Littlefield, 2003.

———. *The Spitting Image: Myth, Memory, and the Legacy of Vietnam*. New York: New York University Press, 1998.

Levinson, Nan. *War Is Not a Game: The New Antiwar Soldiers and the Movement they Built*. New Brunswick: Rutgers University Press, 2014.

Lewes, James. *Protest and Survive: Underground GI Newspapers during the Vietnam War*. New York: Praeger, 2003.

Lifton, Robert Jay. *Thought Reform and the Psychology of Totalism: A Study of Brainwashing in China*. New York: W. W. Norton, 1961.

Luce, Don, "The Tiger Cages of Viet Nam," www.historiansagainstwar.org/resources/torture/luce.html.

Lydon, Christopher. "Camp Termed Showplace," *New York Times*, December 29, 1970.

Lynd, Staughton, and Tom Hayden. *The Other Side*. New York: New American Library, 1966.

Malo, Jean-Jacques, and Tony Williams. *Vietnam War Films*. Jefferson, NC: McFarland & Company, 1994.

May, Elaine Tyler. *Fortress America: How We Embraced Fear and Abandoned Democracy*. New York: Basic Books, 2017.

McCain, John. *Faith of My Fathers*. With Mark Salter. New York: Random House, 1999.

McCollum, L. C. *History and Rhymes of the Lost Battalion*. N.p.: n.p., 1919.

McDaniel, Eugene B. *Before Honor: One Man's Spiritual Journey into the Darkness of a Communist Prison (Before Honor Is Humility— Proverbs 18:12)*. New York: A. J. Holman, 1975.

McEldowney, Carol Cohen. *Hanoi Journal, 1967*. Edited by Suzanne Kelley McCormack and Elizabeth R. Mock. Amherst: University of Massachusetts Press, 2007.

New York Times, "Laird Discounts P.O.W. Interviews," December 29, 1970.

New York Times, "Ex-P.O.W.s Cheer Nixon," May 23, 1970.

New York Times, "400 Ex-P.O.W.s are Given $400,000 Dallas Reception," June 2, 1970.

Nordheimer, Jon. "Postwar Shock Besets Ex-G.I.s," *New York Times*, August 21, 1972

O'Connor, Michael. *MiG Killers of Yankee Station*. Friendship, WI: New Past Press, 2003.

Parsons, David L. *Dangerous Grounds: Antiwar Coffeehouses and Military Dissent*. Chapel Hill: University of North Carolina Press, 2017.

Risner, Robinson. *The Passing of the Night: My Seven Years as a Prisoner of the North Vietnamese*. New York: Random House, 1973.

Rochester, Stuart, and Frederick Kiley. *Honor Bound: American Prisoners of War in Southeast Asia, 1961–1973*. Annapolis, MD: Naval Institute Press, 1999.

Ronco, Theo. "How American Pilots in North Vietnam Lived," *L'Humanité*, November 5, 1970.

Rowan, Stephen A. *They Wouldn't Let Us Die: The Prisoners of War Tell Their Story*. Middle Village, NY: Jonathan David Publishers, 1973.

Ruane, Michael E. "Traitors or Patriots? Eight Vietnam POWs Were Charged with Collaborating with the Enemy." *Washington Post*, September 22, 2017.

Salisbury, Harrison. *Behind the Lines: Hanoi*. New York: Harper & Row, 1967.

———. "A Visitor to Hanoi Inspects Damage Laid to U.S. Raids." *New York Times*, December 25, 1966.

Scarry, Elaine. *The Body in Pain: The Making and Unmaking of the World*. New York: Oxford University Press, 1985.

Silber, Irwin. *Folksinger's Wordbook*. N/A: Oak Publications, 1973.

Sirota, David. *Back to Our Future: How the 1980s Explains the World We Live In Now—Our Culture, Our Politics, Our Everything*. New York: Ballantine Books, 2011.

Smith, George E. *P.O.W.: Two Years with the Viet Cong*. Berkeley, CA: Ramparts Press, 1971.

Solis, Lieutenant Colonel Gary D. *U.S. Marine Corps, Marines and the Military Law in Vietnam: Trial by Fire*. Washington, DC: U.S. Marine Corps History and Museums Division, Headquarters, 1989.

Sonnez, Felicia. "Donald Trump on John McCain in 1999: 'Does being captured make you a hero?'" *Washington Post*, August 7, 2018.

Stockdale, Jim, and Sybil Stockdale. *In Love and War*. Rev. 1984 edition. Annapolis, MD: Naval Institute Press, 1990.

Strong, Pauline Turner. *Captive Selves, Captivating Others: The Politics and Poetics of Colonial American Captivity Narratives*. Boulder, CO: Westview Press, 1999.

Verrengia, Joseph B. "Some Iraq Veterans Find Forgetting the Hardest Part About Killing," Associated Press, April 18, 2003.

Whitehouse, Arch. *Heroes and Legends of World War I*. New York: Doubleday, 1964.

Wyatt, Barbara Powers, ed. *We Came Home*. Toluca Lake, CA: P.O.W. Publications, 1977.

Zaretsky, Natasha. *No Direction Home: The American Family and the Fear of National Decline*. Chapel Hill: University of North Carolina Press, 2007.

Zimmerman, Bill. *Troublemaker: A Memoir from the Front Lines of the Sixties*. New York: Doubleday, 2011.

Oral History Interviews

Conducted by Tom Wilber

Alvarez, Everett. April 26, 2019, Bethesda, Maryland
Beyer, Bruce. October 21, 2017, Arlington, Virginia
Bui Bac Van. January 12, 2015, Vinh, Vietnam
Burchett, George. March 4, 2018, and May 7, 2019, Hanoi
Butler, Phil. April 2018, telephone
Cash, Roy. March 8 and March 24, 2016, telephone
Chenoweth, Robert. June 26, 2017, telephone; November 29, 2017, Hanoi; February 22, 2018, Spokane, Washington; July 4, 2019, Hanoi
Chu Chi Thanh. October 29, 2016, Hanoi
Cunningham, Randy. October 4, 2018, Hanoi
Dodson, Max. February 11, 2017, Golden, Colorado
Dunleavy, Richard. January 28, 2016, telephone
Fant, Robert. March 18, 2016, telephone
Findley, Joseph. February 29, 2016, Maclean, Virginia
Gartley, Mark. February 19, 2019, telephone
Guenther, Lynn. March 9, 2018, telephone
Ha Quang Hung. August 12, 2016, Ho Chi Minh City, Vietnam
Jackson, Chuck. January 16, 2017, Hanoi; April 23, 2019, telephone
Kernan, Joe. September 24, 2018, telephone
Kuwatch, Scott. April 11, 2016, Middletown, Ohio
Le Do Huy. September 20, 2017, Hanoi
Le Khai. March 6 and May 2, 2018, Hanoi
Luu Van Hop. March 8, 2017, Hanoi
Maclear, Michael. March 22, 2016, telephone
Manlove, Don. August 11, 2015, telephone

Mather, Keith. March 8, 2018, Hanoi; August 16, 2019, Spokane, Washington
Miller, Edison. April 24, 2016, and July 5, 2018, Irvine, California
Nguyen Ba De. November 27, 2017, and March 4, 2018, Hanoi
Nguyen Bieu. September 19, 2017, and May 1, 2018, Hanoi
Nguyen Cong Thanh. November 14, 2014, Vinh, Vietnam
Nguyen Minh Y. April 4, 2017; January 29 and July 20, 2018; January 16 and 18, May 6, July 4, and December 4, 2019, Hanoi
Nguyen Phong Nga. February 3, 2017, New York, New York
Nguyen Su. November 12, 2015. January 21, 2017, and December 5, 2019, Hanoi
Nguyen Sy Hung. March 28, 2016, Hanoi
Nguyen Tam Chien. May 2, 2018, Hanoi
Nguyen Thanh Quy. October 4, 2018, Hanoi
Nguyen Thi Binh. November 30, 2017, Hanoi
Nguyen Thi Binh A. January 25, 2018, Hanoi
Nguyen Thi Binh B. January 25, 2018, Hanoi
Nguyen Van Thu. May 16, 2015, Hanoi
Nguyen Van Coc. January 24 and 28, 2018, Hanoi
Nguyen Van Huynh. January 17, 2019, Hanoi
Nguyen Van Ninh. July 3, 2019, Hanoi
Nguyen Viet Bang. November 12, 2014, Vinh, Vietnam
Pham Phu Thai. March 26 and August 15, 2016, Hanoi
Pham Thi Vien. January 24, 2018, Hanoi
Rothstein, Vivian. January 7, 2019, telephone
Ryan, John. February 27, 2017, Morris Plains, New Jersey
Searcy, Chuck. June 13 and November 30, 2017; March 8, 2018, December 7, 2019, Hanoi
Tạ Quoc Bao. May 18, 2016, Hanoi
Tong Tran Hoi. July 20, 2018, Hanoi
Tran Thi Dien Hong. August 12, 2016, Ho Chi Minh City, Vietnam
Tran Trọng Duyet. May 17, 2016, Haiphong, Vietnam; January 17, April 5, and November 27, 2017, and July 2, 2019, Hanoi
Truong Sinh. January 29, 2018, Hanoi
Vo Dien Bien. March 4, 2018, Hanoi
Weiss, Cora. October 17, 2017, December 17, 2018, March 11, 2019, New York, New York
Weiss, Peter. February 14, 2019, New York, New York
Zimmerman, Bill. April 12, 2019, telephone

Filmography

MOVIES

The Bamboo Prison	1954
Blood of Ghastly Horror	1967
The Bridge on the River Kwai	1957
The Bridges at Toko-Ri	1954
Cat on a Hot Tin Roof	1958
Coming Home	1978
Dragonfly Squadron	1954
First Blood	1982
Fixed Bayonets	1951
The Fog of War	2003
The Forgotten Man	1971
Good Guys Wear Black	1978
The Hanoi Hilton	1987
Jarhead	2005
The Lost Command	1966
Madame Curie	1943
The Manchurian Candidate	1962
Motorpsycho	1965
Mr. Majestyk	1973
Mrs. Miniver	1942
The Rack	1956
Rambo: First Blood Part II	1985
Rolling Thunder	1977
Ruckus	1980

Sir! No Sir!	2005
Some Kind of Hero	1982
Stalag 17	1953
Time Limit	1957
Uncommon Valor	1983
The Vietnam War	2017
Welcome Home Johnny Bristol	1971
When Hell Was in Session	1979
The Wild One	1953

TELEVISION

Hogan's Heroes
Jane Fonda in Five Acts
Lassie
South Park
Team America

Archival Collections

Schlesinger Library on the History of Women in America, Radcliffe Institute for Advanced Study, Harvard University
Swarthmore College Peace Collection, Swarthmore College
University Archives and Special Collections, Joseph P. Healey Library, University of Massachusetts Boston
Vanderbilt Television News Archive, Jean and Alexander Heard Libraries, Vanderbilt University
The Vietnam Center & Sam Johnson Vietnam Archive (VNCA), Texas Tech University
The Vietnam-Era Prisoner-of-War/Missing-in-Action Database, Library of Congress
Veterans History Project, American Folklife Center, Library of Congress

Index